"Thompson manages to integrate multiple perspectives, demonstrating how children respond to and internalize their mother's depression." —*Library Journal*

"Tracy Thompson has an uncanny ability to get at the experiential reality of depression. In this vital book, she achieves the same level of nuanced insight that made *The Beast* so compelling and shows how maternal depression consumes the lives of mothers and their children. She looks at the guilt, the anger, the feelings of inadequacy, and the fear that characterize these relationships. Some women may benefit from the coping mechanisms she enumerates; for others, simply the revelation that they are not suffering alone may be sufficient balm."
—Andrew Solomon, author of *The Noonday Demon*

"A moving and deeply personal account of this all-too-common and often ignored phenomenon. Thompson, a wonderful writer, shows how maternal depression can be managed and even overcome."
—Ann Crittenden, author of *The Price of Motherhood*

"Much is now known about depression in mothers and how the illness affects the children and travels through the generations. Modern epidemiology, genetics, and neuroscience are beginning to unpack the risks and causes. Tracy Thompson, a journalist and a mother who suffered from depression, has accessed this information and made it available to the public in a highly readable, current, and humane book." —Myrna Weissman, Ph.D., professor of psychiatry and epidemiology, Columbia University and the New York State Psychiatric Institute

"Writing with tremendous skill and sensitivity from her own experience as a depression sufferer and the daughter of a depressed mother, Thompson uses original research to clarify the relationship between mothers' depression and children's behavior and development. Her message—that depression can be treated and that the process of managing maternal depression can help women become better mothers—will be welcomed by millions of women who have struggled with this challenging disease."
—Judith Stadtman Tucker, editor, *The Mothers Movement Online*

"The book . . . is among the few to address depression in mothers beyond the early postpartum stage—and to talk about what it means for children."
—*USA Today*

About the Author

TRACY THOMPSON is a freelance journalist and the author of *The Beast: A Journey Through Depression.* She lives in suburban Washington, D.C., with her husband and two children.

Also by Tracy Thompson

The Beast: A Journey Through Depression

The Ghost in the House

Real Mothers Talk
About Maternal Depression,
Raising Children, and How They Cope

Tracy Thompson

HARPER

NEW YORK • LONDON • TORONTO • SYDNEY

HARPER

Excerpts from "Continuing to Live" and "This Be the Verse" from *Collected Poems* by Philip Larkin. Copyright © 1988, 2003 by the Estate of Philip Larkin. Reprinted by permission of Farrar, Straus and Giroux, LLC.

A hardcover edition of this book was published in 2006 by HarperCollins Publishers.

FIRST HARPER PAPERBACK PUBLISHED 2007.

Designed by Jaime Putorti

The Library of Congress has catalogued the hardcover edition as follows:
   Thompson, Tracy.
     The ghost in the house : motherhood, raising children, and struggling with depression / Tracy Thompson.—1st ed.
      p. cm.
   Includes bibliographical references and index.
   ISBN-10: 0-06-084379-9
   ISBN-13: 978-0-06-084379-3
   1. Postpartum depression. 2. Mothers—Mental health. 3. Depression in women. 4. Mother and child. I. Title.
   RG852.T46   2006
   618.7'6—dc22   2005054547

ISBN: 978-0-06-084380-9 (pbk.)
ISBN-10: 0-06-084380-2 (pbk.)

07 08 09 10 11 ID/RRD 10 9 8 7 6 5 4 3 2 1

To my mother,
Enley Ruth Buchanan,
June 30, 1926–December 8, 2005

ACKNOWLEDGMENTS

My deepest thanks go to Dr. Sherryl Goodman at Emory University, without whose patient expertise this book would have been impossible to write. Chance brought us together—I found her name early on while scanning titles of books on children and depression—but shared interest in this subject became a partnership, and the partnership became a valued friendship. It was she who designed the survey on which this book is based, she was the person I turned to for advice in helping me puzzle out what the numbers meant, and her comments on some of the early drafts were extremely helpful. The results of the survey were so voluminous and detailed that much of the information remains to be mined. I hope this collaboration will be as fruitful for Sherryl's future writing purposes as it has already been for mine. I hope, too, that the research we have done here will help a lot of women.

Many of the hundreds of women who generously gave of their time and who told me their stories expressed the same hope—that by sharing their experiences, they might ease another woman's struggle. This instinct women so often have to pass along their knowledge probably has a perfectly sensible evolutionary origin, but to me it is one of the marvels of nature.

My editor, Gail Winston, inherited this book at a crucial stage. She restored my flagging self-confidence, gave me quick, insightful feedback that was unsparing, honest, and yet always encouraging, and pulled this project out of the ditch when things were at a low point. I am deeply indebted to her.

I owe an even greater debt to my agent, Beth Vesel, who for twelve years now has held my hand, helped me think things through, and peeled me off the ceiling when things went wrong. Throughout, she has never lost faith in me. I don't know what I've done to deserve her.

Dawn Raffel at *O: The Oprah Magazine* generously helped me get access to precious editorial real estate in her publication to get out an author's query, which was crucial to our survey. My thanks also go to my dear friend Ann McIntire, a meticulous researcher who proved capable of digging up the most obscure articles with amazing ease. Terrika Barham worked tirelessly on analyzing computer spreadsheets and never once complained. Tanessa Sergeant worked equally long hours doing data entry, a truly thankless task.

Bill Kovach suggested early on that the subject of this book might be worthy of a grant proposal—a suggestion that did not end up with my winning any grants but did make me stop talking and start actually doing research. Devra Renner provided me with advice and insight, both in her professional role as a social worker and writer, and in a personal role as a friend. There were others who generously gave their time to look at portions of the manuscript and shared their expertise and insights, notably Karyn Marcus, Sarah Jackson Han, Paul Plotsky, and Ann Crittenden. Others just offered encouragement, including the Girls Night Out Posse, Judith Stadtman

Tucker, Minerva Cox, and Dick Stetler. It is impossible to list all the people in academia who helped to guide me in my research, but Paul Plotsky, Myrna Weissman, Wayne Drevets, Michael O'Hara, and Kenneth Kendler deserve special thanks for putting up with my incessant e-mails and queries—as does, of course, Sherryl.

Most of all, I thank my husband, David, who steadfastly kept on bringing home the bacon during those long periods when I couldn't seem to make any money, who shooed the kids away from my study door when I was guiltily hiding behind it, who fed them dinner and read them bedtime stories, and who has always supported my work, both paid and unpaid, with an innate sense of equality and fairness. He is my first reader, closest friend, and life companion. Finally, I thank my daughters, who were—and always will be—my greatest gifts. They are my constant inspiration, always, to Do Better.

CONTENTS

Motherhood and depression are two countries with a long common border. The terrain is chilly and inhospitable, and when mothers speak of it at all, it is usually in guarded terms, or in euphemisms. I remember the mysterious "nervous breakdowns" my mother's friends sometimes had during my childhood, and the "mother's little helpers" pills so many women seemed to need in order to make it through the day, and I remember the mothers of friends whose daily martini hour started promptly at five. Today, casual drug use is not politically correct anymore, so mothers are more apt to say that they are always tired, or always irritable, or time-starved, or stressed out, or perpetually premenstrual . . . anything, any phrase or word, except *depressed*. Depression is a subject that has come out of the closet, except when it applies to motherhood. Somehow, juxtaposing *depression* and *motherhood*—to raise the possibility that there are some things about the work of raising children that may be stressful, or even conducive to illness—seems bad manners. Motherhood is supposed to make women happy, period.

Maternal depression is a subject I approach from three perspectives: as a person who has suffered from depression, as a journalist with a long interest in exploring the emotional and scientific territory of mood disorders, and—most important, for my purposes here—as a mother.

My own experience with depression began at the age of roughly thirteen. Any number of things might have triggered it, since there's evidence of depression on both sides of my family, but as it happened, it was a car accident and the scars it left on my face that sent me into the hole. The first time wasn't terrible, and I got over it, but there was a next time, and a time after that, and each time was a little bit worse than the one before. I spent the next twenty years trying not to admit there was a problem, trying to outrun it, looking for psychic wound dressings in the form of relationships or workaholism or drugs. I could not admit to myself that I had developed a mental illness until I heard the click of the lock on the door to the psychiatric ward at Georgetown University Hospital in February 1990. I was on the wrong side of that door, very much against my will—an acutely suicidal single woman of thirty-four whose "normal" facade had finally cracked wide open.

Several years later, I wrote a memoir of my experience of coming to terms with this illness. I wrote it as a person who had managed to fairly effectively conceal my problem from myself and others, out of shame, for many years. I was trying to say: *This illness is not what you think. You can be dying and still work a fifty-hour week.* The book was later translated into three languages and anthologized, which told me that it had helped a lot of people put a name to their own problem—one that perhaps they had, as I had, stereotyped as a narrow set of symptoms, or as the kind of thing that happened only to tortured geniuses or street people. After that, I thought, *I'm done with this subject now; I can write about other things.*

And I did. But what eventually brought me back was a life change I had thought of as completely unrelated: becoming a mother. Before I was a mother, depression was merely an illness, something I had learned to manage the way people with high blood pressure or diabetes learn to manage their conditions. I didn't always manage it effectively, but its impact on my life was confined to me and my husband, and there were long periods of remission when I felt fine. At its worst, depression could derail my work, but between 1990 and 1996, that happened only twice: once when I was hospitalized, and another two-week period four years later. If anything, work was a useful distraction during times of feeling bad. It helped hold me together.

But becoming a mother is a profound transformation. It blurs the boundary between self and child, this new being so recently of your body and, for a while, so utterly dependent on it. And even as that dependence decreased and my child grew, and she was joined by a sibling, it began to dawn on me that something else had blurred: the relationship between my illness and my work. Because motherhood *was* work, and I was a mother; my new work was now *me*. Children do not willingly settle for a piece of you; they want *you*, the whole deal, including those parts of yourself that you think of as ugly or which you would give anything to hide. My new job was not something I could use to distract myself during times of depression. Now my depression had a direct and daily impact on my children—may have begun affecting them even before they were born—and this had potentially profound consequences.

At the same time, motherhood brought me a new closeness with my own mother—who, as it happened, suffered a debilitating stroke the year before my first child was born. In the process of closing my mother's house and moving her to an assisted living facility, I went through piles of letters, pictures, boxes of old schoolwork. Memories that I had long ago stored away came to the surface, and elements of my family history that I had not known or had seemed unimportant began to take on new meaning. As pieces of my mother's life began coming into focus, I began to really see her struggle for the first time, and to understand how much my life had been shaped by it. My perception of my own illness began to change. The family myth was wrong: I was not the unstable or wild one of the family, but part of something that began to look very much like a pattern. I saw it in my mother; I heard it in the wistful words of one of my mother's old friends, who told me, "Ruth was never the same" after my sister and I were born. I found it in the descriptions of daily life written by my mother's sister when they were both new mothers fifty years ago. I recognized it in my father's mother, who had spent the better part of my father's childhood secluded in her bedroom after her youngest child was killed in a freak household accident. This was in the 1930s, the era of eugenics, when people with mental illness could find themselves institutionalized or sterilized or simply unemployable. As in many families, depression in my family had been hiding in plain view all along.

And I remembered how, as children do, I had picked up on things adults thought I did not know: by reading a tone of voice, by sensing emotional distance, by becoming the victim of emotions my childish narcissism made me think I had caused, by inventing stories so that these things would make sense. I saw this with a kind of double vision—as child, as mother. And I realized my own children were probably doing the same things.

If I wanted to break this generational pattern, I realized, I had to know what I was dealing with. Eventually, I began to think of what I was looking at as a kind of depression that nobody had adequately described. I began to call it maternal depression.

CHAPTER ONE

What Is Maternal Depression?

Depression is the parent with the most power.

—PATRICIA E. FRANKLIN, FORTY-SIX, OF STAFFORD, VIRGINIA,

MOTHER OF THREE, GRANDMOTHER OF ONE

There is an old red vinyl recliner in the den of the house I grew up in, and it is this chair that I remember her in. She is young in this memory, somewhat plumper than in those honeymoon pictures, but slender still. She wears glasses now, and her red hair, which is beginning to lose its luster, is pinned up severely off her forehead. I am folded up against her, my head in the warm hollow between her arm and breast.

"Flesh of my flesh," she says softly, talking to herself. She takes my hand in hers and turns it over, traces the blue vein beneath the skin on my wrist. "Bone of my bone." And in the wordless way that children know these things, I know she is sad.

Later, when I was older, I would come right out and ask. "What's wrong, Mama?"

"Nothing," she would say, her voice tight as a fist. She would not look at me.

"No, really, Mama. Tell me. What's wrong?"

"I already told you, Tace. None of your business." *Leave me alone.* So I did. And the free-floating anxiety of those unanswered questions wove itself into my earliest memories of her, along with the primitive body-memory of warmth and love that she gave in generous measure. There is no separating the two; they are warp and woof of the same fabric.

Depression is a fire in the brain that, once ignited, is hard to extinguish. It develops its own rhythm of flare-ups and remissions; for many of us, it becomes a chronic, lifelong condition. Motherhood is another lifelong condition: once a mother, always a mother. Which makes it strange that the term *maternal depression* is commonly taken to mean depression in women who have recently given birth—as if depression and motherhood could coexist only during those first few months.

So what is maternal depression—simply Depression + Motherhood? No. It's what happens when a mother's depression reaches out to ensnare her child. It's depression created or exacerbated by stresses common to motherhood, and—most important—it can be transmitted from mother to child via learned behavior, environment, genetics, or any combination of the three. Does being a mother and suffering from depression mean that you are a bad mother? It can, but it doesn't have to. A mother whose child-rearing days are behind her may suffer from depression, but her adult offspring are no longer the captive audience to her illness that they would have been as children. More important, a mother can suffer from depression, recognize it for what it is, and cope with it in ways that spare her children from its effects. In some ways, surviving depression and learning from it can make you a *better* mother.

We do not know exactly how many women in this country suffer from depression who are simultaneously engaged in the strenuous work of child-rearing. But in our culture, women are the primary caregivers for children, and we know that women make up roughly 12 million of the 19 million Americans affected by depression every year (a number that includes 2 million children). A few other statistics help give dimension to the problem. We know, for instance, that women suffer from depression at somewhere between one and a half and two times the rate of men. About one woman in every eight can expect to develop clinical depression during her lifetime. The forms it takes can range from a low-grade dysthymia that lingers for years, eroding quality of life the way rust eats metal, to the mental hurricane known as major depression. Finally, we know that the incidence of depression in women peaks between the ages of twenty-five and forty-four, which are, not coincidentally, the major childbearing years.

Now let us assume, somewhat conservatively, that roughly one-third of the 12 million women currently affected by depression in the United States this year have children at home. That's *4 million* women who get out of bed every morning to face the daunting job of parenting while suffering with an illness that is at best debilitating, and at worst life-threatening. Let us also assume, very conservatively, that each of those women has 1.5 children. That's 6 million children.

Do these numbers seem inflated? It wouldn't be surprising if they did. After all, breast cancer will strike only 215,000 women this year, and it's in the news frequently. But cultural awareness doesn't have much to do with reality; actual risk can be dwarfed by perceived risk. Heart disease, for instance, kills many more women than breast cancer, but women tend to be more terrified of breast cancer than of a heart attack. Moreover, "normal" is not the same thing as "healthy." Consider infant mortality rates: what was "normal" in this country in 1900 would seem shocking today. So it is with maternal depression: if it's not a huge topic, perhaps this is because many mothers simply consider its classic constellation of symptoms—chronic exhaustion and/or trouble with sleeping, dysfunctional eating patterns, low libido, anxiety, loss of pleasure in life, constant feelings of guilt, an inability to concentrate—to be "normal." Ten years ago, more than half the women surveyed by the National Mental Health Association said they considered depression a normal part of aging, menopause, and the postpartum period, and more recent studies indicate that fewer than half the women who experience depression ever seek medical help. As one mother who suffers depression wrote to me, "We don't recognize it because we are inside it." Depression is so inextricably woven into the day-to-day experience of motherhood that simply recognizing it requires a cognitive leap as huge, and as basic, as a baby's first realization that she is separate from her mother.

Adding to the silence is the ever-present stigma of mental illness—and, for mothers, the formidable stigma of admitting to a problem that might make them appear inadequate caregivers. American society (and other mothers in particular) can be harsh in its judgment of mothers who do not live up to the mythical ideal. "I was stunned to discover that many other mothers were also isolating themselves, questioning their personal value and struggling with diminished cognitive capacity," another mother wrote to me, describing her own battle with depression. "We each suffer[ed] alone."

Those letters were just two of the hundreds of stories I began collecting in the fall of 2003, when I began research for this book. They were gathered from queries I placed in 170 newspapers across the country, as well as from a query that ran in the May 2004 issue of *O: The Oprah Magazine*, asking mothers who had had experience with depression to write to me about their experiences, and about how their illness had affected their parenting. At five hundred responses, I stopped counting, because they kept coming in. Some were just polite notes from women who said they would consent to an interview, but roughly three-fourths of them included long and frequently eloquent descriptions of what it was like to live with this illness and be a mother. After whittling down the responses to include only those women who said they had received a medical diagnosis of major depression, I was left with a pool of 393 women. These women were given a 170-question survey designed by my survey collaborator, Dr. Sherryl Goodman, a psychology professor at Emory University who has done extensive clinical research on mothers with depression. The survey yielded thousands of responses, which were analyzed and coded, both by me and by research assistants, in order to validate my basic conclusions. In addition, I did thirty-two in-depth personal interviews—some in extended telephone conversations with women who had written to me, some in face-to-face interviews with women around the country. I met with researchers in the fields of genetics, epidemiology, psychiatry, endocrinology, obstetrics, and brain imaging; I picked their brains on the phone and by e-mail. I spent months reading the medical literature on the subject of women and depression. And I drew from hundreds of e-mails, other personal correspondence, and, of course, my own experience.

My results are limited, in the sense that these women who responded to my queries are a group whose demographics, not surprisingly, largely reflected the demographics of O: The Oprah Magazine readers—suburban, married (three-fourths of them, anyway), educated, affluent but not wealthy. They were women for whom this subject was clearly important: the response rate to the very lengthy survey we sent out was in the 80 percent range. (A 50 percent response rate is usually considered excellent.) Our subjects were, for obvious reasons, a self-selected group, and what they had to say does not, for the most part, reflect the reality of mothers who cope with this illness along with the crushing burden of living in poverty, without medical insurance. It is sobering to realize that in many ways the pictures that emerge here are of women in relatively privileged circumstances.

There is an important difference between research on depression involving women who happen to be mothers and research aimed at defining what maternal depression is. There is a large and growing body of the former in the medical literature today—studies that track the percentage of women who suffer depression during pregnancy, for example, or measure physiological phenomena that are associated with depression, or epidemiological research tracking the incidence of depression across generations. That kind of research is quantitative. It uses standard research tools: numerical scores on psychological inventories, salivary cortisol levels, clinical assessments of social functioning, comparisons of data from depressed mothers with data from a control group. In the survey, Sherryl and I used some of those tools, too, in order to measure the women's responses against some kind of objective yardstick that would indicate the actual incidence of depressive symptoms in the women and their children, and the effect of family discord on the children in the household. But for the most part, the information we sought in the survey was information that could not be found by using quantitative measurements. It was *descriptive*, or qualitative. The premise was simple: if you want to describe maternal depression, you begin by talking to mothers who are depressed.

So I did. Most of the stories I heard illustrated what it's like to deal with this illness as members of the middle class, with health insurance and ready access to doctors—conditions that, as it turns out, provide no guarantee of good medical care. A few women talked about what it's like to deal with this illness and to be poor. Older women told me about the way it used to be, back when a wife who suffered from severe depression in every pregnancy still needed her husband's permission to have a hysterectomy—all of forty years ago.

The responses fell into three broad and interrelated categories describing patterns of behavior—reactions that I viewed not as examples of "bad mothering," but simply as symptoms of illness. The first was a depressed mother's habit of withdrawing—emotionally, physically, or both—from interaction with her child. Some mothers just went to bed and slept for hours. Others said they mechanically went through the motions of daily life with little or no emotional involvement. Some kept up a frantic round of activities that keeps real conversation to a minimum, got obsessive about house-work, or plied their child with diversions. "When I am de-pressed I will do whatever I need to do to get her taken care of and out of my way," one mother wrote. "Later I feel so guilty. She senses that there is something wrong, but isn't really sure what it is." The pull toward withdrawal was so strong that many mothers said they aggressively defended it. One mother confessed that she would sometimes tell her child, "If you wake me again, I'll smack you."

In that way, withdrawal was connected to the second pat-tern of behavior: chronic hyperirritability. The mothers who an-swered my questions described, for instance, yelling, "Shut up!" to a child humming absentmindedly to himself in the backseat, or responding to a toddler banging his spoon against his cereal bowl by snatching the bowl away and yanking the child from his seat. They had the kind of anger people feel when they have too many things to do, too little time to do them in, and con-stant interruptions. Their simmering irritation leaked out in the form of snappish responses, nagging, demeaning or sarcastic re-marks, or lashing out at a child who did not meet inappropri-ately high expectations. "I just yell all day at them when I'm depressed," one mother wrote in the survey. "I have absolutely no patience whatsoever. It's like I have no coping skills as a par-ent. That just vanishes." Another wrote: "When I'm cranky I don't address a lot of my son's misbehavior until I get to the point where everything he does bothers me. Then when he doesn't listen to me [at] once, I overreact and yell at him for something as small as not getting his cup for me." More than one mother described the jolt of recognition she got when she saw fear in her child's eyes—the same fear she had felt as a child growing up with a depressed mother. It wasn't until then, they said, that they realized they were depressed.

Withdrawal and/or hyperirritability result in a third pattern—the inability or unwillingness to impose limits on behavior. Many of these mothers described a habit of avoiding conflict. A hyperirritable mother may grit her teeth when she sees her child misbehaving because she fears overreacting; a mother who is suffering the physical depletion induced by depression can simply be too tired to make an issue of it; the mother who is upstairs asleep doesn't even know about it. Whatever the reason, the misbehavior goes unaddressed—or, which may be worse, gets addressed only sporadically. Other mothers described allowing their children to treat them with rudeness and disrespect. Still others described how their children took the verbal abuse they dealt out and turned it onto their siblings—or, at times, directed it right back at them.

Any one of these could be a Bad Day scenario most mothers can recognize. But I'm not talking about a bad day, or even one of those bad patches every family goes through from time to time. Maternal depression is a Bad Day that comes for a visit and refuses to leave.

Maternal depression can also be described as a constellation of behaviors that are a reaction to a very specific stress: the demands of children. The whiny toddler gets a snappish "Shut up!" instead of the nap he needs; the new mother with the wailing baby feels resentful or indifferent; the active eight-year-old gets a mommy who would rather sleep than host a playdate; the teenager with the nasty attitude gets a mother who lets him get away with it. The behavior of the mother, in turn, is a powerful influence on the behavior of the child. Some children become caregivers to their mothers, growing up too soon; others simply mimic her responses to life. Only 14 percent of the women in our survey said they thought their illness had not affected their children at all. Of the rest, a striking number used the terms *perfectionist* and *self-critical* to describe their children. It's a mind-set very close to one of the hallmark cognitive distortions of depression, which is all-or-nothing thinking. Simply growing up with a chronically depressed parent can teach a child habits of thinking that deprive him of emotional resiliency and the self-confidence to risk failure. The worst effects of maternal depression occur when there is no buffer, when the messages children pick up from their mother's behavior become part of their unconscious operating assumptions about life. *If I am really good and clean the house, Mommy won't be sick.* Or: *Why would she want to spend any time with me? I'm not worth it.*

Even then, there are psychological buffers that can go a long way to minimizing the impact of maternal depression, and these were described, too: a healthy father, grandparent, or other relative who is able to pick up the slack, a teacher who takes the time to encourage, and—not least—the mother's ability to recognize her illness and get effective treatment. Any one of these—or better yet, all of them—can act as a bulwark around a child's mental health. Children grow like weeds; sometimes, they're just as tough.

This last part of my working definition—maternal depression is depression "which can be transmitted from one generation to another via learned behavior, environment, genetics, or any combination thereof"—has the disadvantage of being a definition that can only be applied retroactively, but that's a common problem in psychiatry. There are still unanswered questions about how antidepressants work, and in common clinical practice, the diagnosis of depression is often the result of a strange backward reasoning that would look absurd if it were applied, say, to orthopedics. You have no energy and don't feel yourself; your doctor gives you a prescription for antidepressants; you feel better. There's your answer! You were depressed. That children of depressed mothers are much more likely to suffer depression themselves—as well as a range of other psychiatric disorders—has been documented for decades, beginning with the pioneering work of Myrna Weissman and Eugene S. Paykel in the early 1970s. Consequently, it was no surprise that 90 percent of the women in this survey said they had a parent who suffered from this illness (more often than not, their mothers). Most of the cutting-edge research in this field is exploring how genetic vulnerability intersects with environment—a broad term that encompasses everything from life inside the womb to the psychological stress of divorce.

My aim here is both empirical and journalistic. The first part of this book simply describes what it is like to experience depression in the context of child-rearing. In a way, it's a collection of narratives, one that helps illuminate what day-to-day life looks like and feels like for depressed mothers. How does it affect their parenting skills? What are some of the triggers, and do those triggers always include postpartum depression (PPD)? (The quick answer is no, but PPD is a significant part of the picture.) How do these mothers describe their children? What roles do husbands and partners play?

The second half of the book delves into the science of this

field: research on depression in general that is illuminating the pathways by which this illness travels from one generation to another, and some of the ways in which environment helps or hinders that process. It will describe ways that women have found to cope with this illness—both hurtful and helpful methods. And it will offer words of guidance and hope from women who, despite this illness, have thrived and raised healthy children. Finally, it will speak to the most important question facing mothers with depression: What are the best ways to make sure that your children do not suffer the way that you do? Is it possible to break this generational pattern?

To the last question, I give you an answer right now: yes. The very factors that make depression a complex illness also mean it is possible to attack it from any number of fronts. You can't change your genes—at least, not yet—but you can benefit from the insights of cognitive behavioral therapy. Other kinds of therapy help. Exercise helps. Having a good social support network is vital. Many people find strength in religious faith or serenity in meditation. Not least, we live in an age of drugs that can effectively treat the worst symptoms of depression; in maintenance doses they allow many of us to lead normal lives.

Drugs are not cure-alls; they are tools. But they can be extremely valuable tools, and can jump-start the brain's natural healing response. This is especially valuable for women with maternal responsibilities. Dump any preconceptions you have about antidepressants. If you need them, take them. "Get medication!" wrote one mother. "No extra points for suffering!" Wrote another: "I swear, they should give out T-shirts when you leave the maternity ward at the hospital that say, 'WHATEVER WORKS!'"

In fact, over the last twenty years, the list of what works has become longer, more precise, and increasingly effective. Until the mid-1980s, there were only two classes of drugs to treat depression—tricyclic antidepressants and monoamine oxidase inhibitors (MAOIs), both of which had side effects so unpleasant that only the truly desperate would put up with them: dry mouth, tremors, weight gain, sedation, and, in the case of MAOIs, severe dietary restrictions. In 1987 came Prozac, the first antidepressant to target levels of a specific neurotransmitter, serotonin, without those debilitating side effects. Today there are number of serotonin reuptake inhibitors (SSRIs, like Prozac) on the market, as well as newer drugs targeting the neurotransmitters dopamine and norepinephrine. A third class of medications chemically unrelated to SSRIs, namely Effexor and Wellbutrin, brings the arsenal of antidepressant drugs up to more than twenty, and most of those drugs are now prescribed in combinations tailored to individual patient needs in ways that were impossible even ten years ago. Research has since moved beyond simply manipulating neurotransmitter levels to look for the cause of the dysregulation. That, it turns out, has to do with malfunctions in the body's endocrine system—namely, a hyperactive hypothalamic-pituitary-adrenal gland system (the HPA axis) that oversecretes corticotrophin-releasing hormone (CRH). The latest drug frontier is the development of CRH antagonists, which would slow down the overactive HPA axis.

At the same time, advances in brain imaging techniques, notably the development of positron-emission tomography (PET) scans, have heightened our knowledge of brain anatomy on a microscopic scale in the same dramatic way as the Hubble space telescope has sharpened our perception of the cosmos. Being able to see the brain better has been invaluable in any number of ways, from tracking the effects of various drug treatments to testing the efficacy of various kinds of therapy. By showing researchers, for example, that antidepressants cause the growth of new nerve cells in the hippocampus—a finding that could explain why it often takes weeks for them to have a therapeutic effect—or that prolonged stress and/or depressive episodes result in structural changes in highly specific parts of the brain, the new imaging techniques help researchers avoid blind alleys and point the way to the most promising areas of inquiry.

What might be most important, however, is research into the biological mechanisms by which a vulnerability to depression can be transmitted from mother to child—because discoveries in this area are proving conclusively that genes are not destiny, or even close to it. We are all born with various genetic propensities to one thing or another, whether it is a preference for the color blue, a woodworking hobby, or a tendency to develop clogged arteries. By altering environment, we can affect whether those genes ever get expressed—that is, whether they get "turned on"—and this seems to be the case as well for at least some of the genes that confer a vulnerability to depression. In short, depression may be every bit as amenable to lifestyle changes as heart disease or diabetes. For some women, particularly those with a history of depression, lifestyle changes may begin even before their children are born: research into the effects of untreated depression during pregnancy indicates that the high levels of stress hormones secreted during depressive episodes, mainly cortisol, can harm the fetus's developing HPA axis. That, combined with the fact that certain antidepressants have so far caused no harm to fetal health, points to the possibility that the prevailing obstetrical wisdom—"a drug-free pregnancy is a healthy pregnancy"—may in some cases be wrong. Those findings are preliminary, but published research has shown how environment and genetics can interact in adults: when confronted with major life stress, adults with a particular type of serotonin transporter gene suffer depression at more than double the rate of people who have another type. Knowledge like this can be a road map for mothers with depression, helping them to foresee, and possibly forestall, things that can trigger an episode in themselves, as well as eventually helping their children understand their own potential vulnerabilities.

On a cultural level, things are changing, too. Slowly, doctors are beginning to understand that they have been overlooking this illness in their patients; even more slowly, but moving in the right direction, health insurers are coming to the realization that the brain is an organ, just like the heart or pancreas, and deserving of equal coverage. It used to be that children were thought to be immune from depression; now we know better, and despite scattered reports of increased suicidal behavior in adolescents taking some kinds of antidepressants, it is equally clear that, in many cases, early treatment can make a huge difference.

As a journalist, I've learned that one sign of finding a good story is when, after a long struggle, you figure out the right questions to ask and, suddenly, new information comes pouring out. That was what struck me about the survey responses, letters, and e-mails: their urgency. These women longed to tell their stories, their words tumbling out in the way words do when they have been left too long unspoken.

"I am madder than hell that [mothers with] clinical depression are not treated with the awesome wonder they deserve for doing a very hard job while struggling with a very serious disease," wrote one woman, who wished to remain anonymous. Then she added, "Because of mental illness stigma, I am very careful [who] I tell. I am well aware that I thereby perpetuate the stigma."

"A mother who is not treated for depression is a . . . vacant mother," wrote Beth Celona, of Melrose, Massachusetts. The stakes in this struggle were vividly put by Linda Green, from Pacifica, California: "I thought it was the hardest decision to make: choosing to live when I didn't want to. But I was wrong. *This* is the hardest thing to do—trying to let my daughter find out for herself that her own life is worth living." She was in company with the New Jersey mother who has suffered from depression for years, who wrote of watching her sixteen-year-old daughter get her stomach pumped after an aspirin overdose and realizing that "the poison of depression had to stop here."

Another mother described the hard-won blessings of this disease: "I have seen myself as completely broken, and then accepted and loved even in that broken state. This is a powerful peace and sense of grace that I can now teach my own children." Then there was this, from a fifty-seven-year-old woman whose own mother suffered from depression, and who has struggled with the illness herself since her early twenties. She raised her own daughter with the determination "to be the parent I did not have." Now her daughter is sixteen.

"I feel sad for the 'hero' in me because nobody knows how hard it's been—or how rewarded I feel when I look at who she is," she wrote. "I feel like someone who has climbed Mount Everest. When I look at her, it's like sticking my victory flag in the snow."

It was an oddly familiar metaphor. I have a recurring dream about climbing, too, my progress always painful and slow, the climb getting steeper with each step. Depression is like a mountain standing in your way, and the path over it is risky. But for those who persevere, who get a few tools, a little luck, and a lot of help, it is a journey that gives you a view you won't get anywhere else.

CHAPTER TWO

Motherhood:
The High-Stress Occupation

The worker can unionize, go out on strike; mothers are divided from each other in homes, tied to their children by compassionate bonds; our wildcat strikes have most often taken the form of physical or mental breakdown.
—ADRIENNE RICH, *Of Woman Born*

*M*y feeling for my children does not surpass my desire to be free of their demands upon my emotions," poet Anne Sexton once told her psychiatrist. "What have I got? Who would want to live feeling that way?" She killed herself at forty-six—a great poet, a woman who suffered from depression and, at least according to prevailing cultural standards, a really rotten mother.

But Sexton told the truth. Children *are* inconvenient, and the more beloved they are, the more inconvenient they get. There is no end to their demands, and for mothers there is no end to the guilty sense that at any given moment, some need of theirs is not being met. As I write these words, it is night, and I am stealing work time from what should be my daughter's bedtime ritual. She comes into my study: "Mom, you *promised*." My work needs me; she needs me. She needs to talk right now; I am fighting the daily battle to carve out hours to write. But do I *need* to be a writer? *You're selfish; if your work is not somehow providing for your children's necessities or their life enrichment, you're just massaging your ego,* says a voice in my head. Then comes another: *No! Women are more than mommies; don't you want your daughters to know this?* This mental point and counterpoint takes a tenth of a second, and is of no interest to my eight-year-old; she is literally getting in my face. "You *promised*."

The day-to-day work of motherhood, whether mothers work in an office or at home, is a cross between the life of an administrative assistant (being at someone else's beck and call) and the life of a midlevel manager (being the person who does most of the day-to-day running of the household). In the salaried workforce, administrative assistant and managerial jobs rank numbers one and two, respectively, in terms of mental stress, according to the National Institute for Occupational Safety and Health. Police work has its adrenaline surges; coal miners are more likely to face physical injury; medical lab technicians face exposure to deadly viruses. But in terms of continuous, daily mental job pressure, motherhood ranks right up there on the stress-o-meter. This fact is crucial to understanding maternal depression. Why? Because of all the discoveries about depression that have emerged in recent years, what stands out the most is that depression is fundamentally a dysfunction of the body's reaction to stress.

Motherstress, the term I coined for this phenomenon, is as ubiquitous as Muzak. It's something we tend to notice only in passing, the way we absentmindedly hum along with James Taylor while cruising the produce aisle. Unlike Muzak, motherstress cannot be toned down or turned off; it's our own personal piped-in radio channel broadcasting 24/7. The underlying cause of all this is that we live in the era of what sociologist Sharon Hays calls "intensive mothering," in which the standards for what constitutes a "good mother" have reached all but unattainable highs. Mothers have been held to high standards before now; Victorian mothers, for example, had some lofty ideals to live up to. But it was a rare Victorian mother who had any other day job. Most of today's mothers do—and even mothers today who "don't work," who are at home all day running their households and raising their children, are doing it in a far more complex society. The bar has been raised in imperceptible increments, for such a long time, that much of the time we don't even realize that we are holding ourselves to standards our mothers never had to meet.

My mother was a fairly typical mother of her generation, as well versed in child care advice as her peers and, because of her own background as an orphan, highly motivated to give her children the happy childhood she had missed. And yet my mother felt no obligation to spend "quality time" with us, except for the purpose of meals and nighttime prayers. When we got in her way, she shooed us out the door to play with the neighborhood kids, whose mothers had done the same with them, and the only rule was "Be home by suppertime." Imposing that rule was being a good mother; any mother who expected to know where her child was at every moment of the day was "hovering" or "being overprotective" and ran the risk of her child turning out to be a mama's boy or (if a girl) a crybaby. This was not the manicured world of some affluent suburbia, either; we lived in a once-rural area that had almost, but not quite, been swallowed up by industrial development. My sister and I used to spend hours exploring the abandoned Armco Steel plant not far from our house, poking sticks into vats of God knows what kind of toxic sludge. My mother made us walk unaccompanied to the bus stop even when it was cold or when it rained. We never went to summer camp, unless you count church camp, which was basically Sunday school in shorts and lasted only a week. To her, a perfectly acceptable child's birthday party was a cake, some candles, and three kids from the neighborhood playing Pin the Tail on the Donkey in the backyard. Total cost: maybe $15 in today's currency, including streamers.

Yet by the standards of her day, she was extraordinarily involved and attentive. For most of my childhood she did not work outside the home. She taught me phonics: it was while sitting on the living-room sofa in the crook of her arm, looking at a book with her, that letters first came together to form a word in my brain. She took us to extracurricular activities; my sister was in 4-H and Junior Achievement, and both of us took piano lessons. She ran Vacation Bible School at church. She tried to get us interested in Girl Scouts but my sister and I were either too lazy, too bookish, or maybe both, to be interested. She saw to it that we got swimming lessons; she took us to the library. This was in Georgia of the 1960s, a conservative environment still very much in the era when women were encouraged in every possible way to leave those wartime jobs and go home, become domestic goddesses, and raise children. The standard for what constituted a "good mother" was already on the rise.

But it was nothing like today. Take, for example, this comment from one of the mothers in the Motherhood and Depression Survey about her toddler son: "I will try to read the newspaper when he is eating. I will try to read a magazine when he is playing. Sometimes I feel guilty about this, thinking he should have all of my attention all of the time." If I had pestered my mother when she was trying to read the newspaper, she would have had two words for me: "Scram, kiddo." So where did this mother get the remarkable idea that her son "should have all of my attention all of the time"? She probably has no idea, just as I couldn't say exactly where I got the idea that when I drove anywhere with my firstborn, I had to babble nonstop. "Look! A big red truck! Can you say 'red'? . . . Oooh, that was a big bump, wasn't it?" Did Rebecca become the highly verbal child that she is because of our little car talks? It's just as likely she would have enjoyed some peace and quiet. I would have.

What Sharon Hays calls the "socially constructed" ideology of intensive mothering is based on a few assumptions that are, in historical terms, quite new. These include the belief that (despite lip service to notions of father-mother equality) mothers are uniquely fitted to fulfill their children's needs; that the fulfillment of those needs, and the development of the proper mother-child bond, is the foundation upon which the child's future mental health depends; and that all this requires a mother's vigilant watchfulness for the child's every waking moment—or, if not that, an extraordinarily high duty to choose only the best possible substitute, to take the mother's place for as brief a time as possible. In a nutshell, it's the belief that "there is something . . . children need that only [mothers] can give them."

By these standards, my mother's performance would rate as barely adequate. Where was the "attachment parenting"? Where was the "floor time," the Mozart in the nursery, the specially designed baby mobiles that stimulate eye-hand coordination? My mother left us to entertain ourselves; if I ever said I was bored, she got out the ironing board and a pile of my father's damp shirts and said, "This should keep you busy." We had no daily schedule of soccer or karate or "social skills" classes or private gymnastics lessons. We never got tennis or ballet lessons. We never belonged to a soccer league. She never signed us up for any kind of summer enrichment camp; the first time I ever heard the phrase "art history," I was in college.

Mothers who would choose such an approach today—assuming they are not poor—would be viewed as being seriously inattentive to their children's mental development. Which is just as well, because in practical terms, it's no longer possible for most of us to take this approach. I can't shoo my kids out the door to play with other kids because there are no other kids out there to play with. Every child in the neighborhood is booked in advance with playdates, museum trips, soccer, Tae Kwon Do, calligraphy classes, ballet lessons—you name it. In the afternoons and during the summer, the streets in my neighborhood are bereft of children except for those who pass by looking out the back windows of minivans, on their way to some appointment. You can see the physical expression of these changing standards by driving through almost any suburban neighborhood: the smaller parks are used for organized soccer practices and the like, while the public swing sets and seesaws rust away from neglect. (At bigger parks, the equipment gets more use—from the buses that periodically disgorge dozens of kids from a local school or day care center.) At the same time, roughly one backyard out of every three boasts a private play castle, or pool, or trampoline, or all three. Parks are public places, intended for spontaneous use; backyards are private, and to enter them you must be invited—which implies planning, selection, and a certain degree of supervision. Guess whose job this is? Meanwhile, the calendar for the local moms' club is so crammed with craft sessions, hints on where to shop, and lists of places to have a learning experience with your child that even a stay-at-home mother has great difficulty staying home.

But that's just the beginning. Mothers today are also often expected to plan household meals, clean the house or hire someone to clean it, keep track of the social lives of various family members, possibly see to the care of elderly parents hundreds of miles away, set up after-school care, provide cupcakes occasionally for classroom parties, file medical insurance claims, sign the report cards, make and keep doctors' appointments, make sure the family computer has up-to-date virus and spyware protection, maybe even pay the monthly bills or monitor investments. The list of things required for the smooth functioning of the family unit is much longer today, and Mom is usually the default option for the person who gets to do them.

What's a normal day for a "nonworking" stay-at-home mother living in an upper-middle-class suburban paradise? "I get up [at 5 A.M.], get dressed, empty the dishwasher [and] get breakfast," said Liz Bunker, a thirty-nine-year-old mother of two elementary-school-aged girls who lives in Peachtree City, Georgia. "Then I get the kids up. One wakes up happy and the other one doesn't. I get them started on getting dressed and fed. If Mark is not going in early, he'll help out." After she drives the kids to school, "I come back, walk the dog, get my own exercise. Some days, I do substitute teaching. But ideally, I do errands after I get my exercise. Then there are chores: gardening, groceries on Friday, cleaning the house on Thursdays. On Tuesday and Wednesday I have Brownie meetings. . . . Then at 2 P.M. I pick up the kids. Then there's snacks, homework, and then usually [a playdate] after school. Then there's dinner, getting the kids bathed, getting their teeth flossed, reading stories to them, prayers. Sometimes I go to bed right after them. Oh yeah, and I do laundry in between all those things."

"We have one of the most egalitarian marriages I know," said another mother, a Presbyterian minister with an entire congregation to tend to—but then she goes on to say that in her household, she's the one who tracks the kids' doctor appointments and keeps the household running. Why? "A mother is still a mother in our society."

The constant juggling of all these tasks creates what I call time fragmentation. The constant demands of children, especially small children, means that no sentence, no thought, no task can be completed without interruption, often by the insertion of an even more important task, which can then be superseded by a task of the utmost urgency. Effectively dealing with time fragmentation is a highly compensated skill in, say, air traffic controllers, but it's just as crucial to managing motherhood. Very early, mothers learn to communicate with each other in staccato bursts of speech, interrupted by a child's request or a pause to sign for a package or the need to referee a playground crisis or any one of a thousand other things. And many deal with this sensory bombardment while suffering the effects of chronic sleep deprivation (for many women, the birth of a child means the end of an uninterrupted night's sleep for the next five or six years).

Yes, you say, mothers do all these things—but is it really all work? It's not work to love one's child, is it? Staying home and running a household is called "not working," and so-called working mothers just juggle—a word with a cheerful, circus-y connotation. And it's true that much of what I have described is done in a loving spirit, for people we love. One of the reasons motherhood brings intense pleasure is that it is a blissful merger of Freud's famous definition of happiness: "To love and to work." Our culture, however, subsumes everything into the first part of that definition. As Sara Ruddick puts it in her book, *Maternal Thinking,* "Mothering can be imbued with such passionate feelings that onlookers, accustomed to distinguishing thought from feeling and work from love, can barely recognize amid the passion either the thinking or the work."

The work that mothers do is like the curvature of the earth: it's there, but you can't see it. Take, for example, this passage from a 1985 collection of oral histories of the rural Midwest in the late nineteenth and early twentieth centuries. One elderly Indiana resident recalled that after his mother got up early, cooked breakfast for the family, and did the dishes, she would take him and his siblings out to the fields, where they played in the shade while she worked alongside her husband. At noon, she cooked lunch and did the dishes while everyone else rested, then returned to the fields to work until dusk. The speaker never questions this unequal division of labor, though the passage does end with the (probably rueful) quote: "I often wondered why mothers didn't get tired."

The invisibility of the work mothers do is partly an example of the Law of Unintended Consequences. Feminists in the early 1900s consciously abandoned an interesting idea that was just beginning to germinate at the time—the notion of according domestic labor some kind of economic value—for strategic reasons: it seemed more important to frame the debate around giving women the freedom to seek careers and interests outside the home. As that dream became a reality, the battle for pay equity in the workplace created pressure on women to perform in the workplace like men in every possible way—a pattern that, for the most part, still holds sway. The unintended legacy of those earlier generations of feminists is the silence that surrounds the subject of unpaid caregiving today, and the uncomplaining acceptance of workplace expectations that shoves child-rearing into the margins of everyday life.

The work of motherhood is also invisible because mothers are conditioned not to talk about it. There is a passage in Rachel Cusk's book *A Life's Work: On Becoming a Mother* about the hideousness of dealing with cranky children on a rainy Saturday morning—the screaming fights over the television, the crunch of cornflakes on the kitchen floor, the sleep-deprived parents' grim search for rainy-day diversions—that my fellow mothers and I think is falling-down funny. But when I read it to my husband, he was appalled. "Tell me it isn't that bad for you," he begged. Other women friends told me similar stories of reading it to their husbands, none of whom thought the passage was amusing. And so I retreat to the safety of what Cusk calls my "coven of co-mothers," where we cackle at our private joke.

But at its most basic level, maternal invisibility is born of the infant's utter dependence—which, assuming his or her needs are met, becomes the blissful assumption that Mommy's breast will be there, Mommy's hands will comfort, Mommy's voice will soothe. Utter reliability breeds invisibility, as every good servant knows. "I know when they'll be hungry, and the food is ready; I know when they'll be tired, and the bed is turned down," says Mrs. Wilson, the perfect servant of the English manor house of Robert Altman's movie *Gosford Park*. "I know it before they know it themselves." You see this phenomenon at work in, say, a house filled with adults and children. The grown-ups are talking, having drinks; the children are running and playing in other rooms. At some point, the hostess looks at the clock: the kids will soon need something to eat. She disappears into the kitchen. Another woman quietly gets up and joins her, or maybe two. In fifteen minutes, the kids have been rounded up and are eating macaroni and cheese. Everybody else—the other adults, the kids—can relax, not even realizing that they are relaxing, oblivious as to how that macaroni and cheese got there (or where in the refrigerator the milk to make it can be found). Meltdown has been averted, the party can go on. The moms made it happen because that is what moms do.

This skill—the ability to exist simultaneously in the moment and half an hour from now—is an important cognitive tool, and just one example of the most invisible maternal work of all: the mental stuff. We recognize intellectual labor when we see a mathematician creating a model for the most efficient delivery system for, say, a General Motors plant—but somehow we don't when a mother brings the same sort of logistical skills to bear on planning a week's schedule involving two kids, three soccer practices, day care drop-offs, one piano lesson, one dentist's appointment, and a trip to the grocery store (subtopic: meal planning). It's hard to see because the mental labor of motherhood does not conform to classic notions of scientific rigor. As with the scientific method, experimentation is the most valued tool—but in the work of motherhood, the goal is to find out *what* works, not how. It assumes that external conditions are fluid. My husband and I were having a heated argument once about how to get Rebecca to do her homework. The problem had gotten severe, her teacher was making pointed remarks about her classroom behavior, the school counselor had gotten involved. I suggested several courses of action and trying all of them immediately. "No," David said. "Because if we try all of this at once we'll never know what works."

"Yeah," I snapped, "and by the time you get the experimental protocol all worked out, she'll be ready for college"—at which point he stalked out of the room. He was thinking like a scientist, with an eye toward developing a plan for dealing with the same problem in the future; I was thinking like a mother, assuming that the problem at hand was unique. "I don't care if I *never* know what worked," I told him later, when things had calmed down. "Whatever works this time might not work the next, or maybe it will be a combination of things, or maybe it won't be of any use with Suzanne, and anyway, by this time next year everything will have changed.

So why do we need to know?" To some extent, I had to admit later, David had a point. But childhood is *now*.

The mental work also involves the ability to make dozens of fine-tuned judgment calls every day. A simple request— "Can I have a snack?"—triggers a dizzying cascade of associated questions. *What time is it? Did this child eat breakfast? What kind of snack do I give her? What is the nutritional advantage, if any, to peanut butter versus an orange, in light of what else she has eaten today?* The questions are complex, the need for mental speed acute. My friend Sarah once was driving a carload of little girls, including her daughter, back from a birthday party in her minivan. The sunroof was open, and when a great gust a wind came, the balloons the little girls were holding were suddenly sucked up and out, swept away in an instant. In the nanosecond interval between that event and the onset of mass hysteria, Sarah said, "Quick! Close your eyes and make a wish!"—they did, every last one, squinching their eyes shut and, without realizing it, converting a catastrophe into a moment of magic. I don't know what to call that ability. *Creativity* doesn't cover it; creativity is often slow and sweat-soaked and agonizing. The closest I can come is that underused word *motherwit,* which my dictionary describes as "natural or practical intelligence, wit, or sense." If it involved computers, we would call it genius.

Why does all this matter? Because it is basic human nature to want some recognition for what we do, especially when there's no paycheck involved. To forever labor in obscurity is demoralizing, even if it's all you ever expect; the only thing worse is to work in obscurity *and* to feel that you'll never really get it right.

So unrealistic cultural expectations, the demands of an increasingly complex society, the inherent difficulty of the work,
combined with lack of social recognition—they all add up to
motherstress. Reactions to stress vary widely and are greatly
influenced by genetics; plenty of mothers deal with all this in
sensible, healthy ways. But it's worth noting that the body's
reaction to low-grade, chronic stress is to secrete stress hormones and to keep those levels elevated. Long-term exposure
to those hormones increases vulnerability to insulin resistance
(often a precursor of diabetes), increases the risk of cardiovascular disease, and depresses the immune system. In the
brain, long-term exposure to stress hormones kills off cells in
the hippocampus, a region of the brain that is central to
memory and learning—an effect that, in turn, has been directly linked to major depression.

In the mornings, I get up at 5:30 to have forty-five minutes—
an hour, if I'm lucky—of quiet time to myself, to read and
savor that first bracing cup of coffee. It's a race, because the
children are early risers, too. I sneak past Rebecca's room like
a burglar and tiptoe through the kitchen to turn on the coffee
machine. David gets up soon after, and in the winter he
builds a fire. My chair faces the window, and I watch as the
blackness outside resolves into the shape of bare branches
against a sky of the palest blue, which gradually pinks up behind the neighbor's house. We read in companionable silence.
But soon enough we hear quick footsteps, and Rebecca
comes in, dragging one of her many blankets. She collapses
on the sofa and curls up as if to go back to sleep. "Hey, Mom,"

she will say, sleepily, and despite the passionate argument about something that will happen between us that day (and there always is one), her greeting is as smooth and cool as river stones. She is my hill-and-dale girl, who came into the world unsettled and howling, who flings herself into my arms for fierce hugs, then delivers some outrageous bit of back talk—"Talk to the hand, 'cause the face ain't home"—when I remind her to pick up her clothes. She is eight now, a child of strong enthusiasms—endearingly naive one moment, unsettlingly sophisticated the next.

Then we hear a soft complaint—Suzanne is having trouble herding all her stuffed animals into the family room with her—and then, after a pause, there is a pattering of footsteps and I set my coffee cup aside just in time for her to hurl herself into my lap. She burrows in, small legs tucked beneath her, head settled close to my heart. Suzanne was a surprise, born after we thought we were out of the baby-making business; now it startles me that I ever could have imagined a life without her. Rebecca is her father's child, with her green eyes and honey-brown hair. Suzanne comes from my end of the gene pool—dark blond curls, skin so pale the veins show through, blue eyes and (unlike me) impossibly long lashes. Once I showed her a picture of me when I was two. "Suzanne!" she said delightedly, pointing at it. "No," I told her. "That's Mommy when she was your age." She scowled. "Suzanne!" she insisted.

Now I bury my face in her hair, catching a sweetish whiff of little girl pee. "Apple juice, Mommy" she says, and I've got my orders; time to start the day.

"Your life is about to change," people told me before Rebecca was born, but they were wrong: our lives were about to

disappear. For most people, the arrival of the first baby is a period of joyful chaos. For David and me, it meant that virtually
everything I had learned up to then about dealing with depression was now either inadequate or useless.

Take, for example, that bedrock rule for dealing with this
illness: *Take care of yourself.* My "self"? New mothers have
two selves: their own body, and the new one so recently and
violently separated from that body. This was something I realized the night we brought Rebecca home from the hospital:
"I" was now, and for the indefinite future would remain, a
"we." In normal circumstances, this is blissful. But if you are
doing a free fall into postpartum depression (PPD), as I was,
it comes as a sickening realization: your plane is going down,
there are two of you, there is one parachute.

We made it through that awful time, but it was a preview of
what was to come: the experience of being a mother who copes
with depression. I had not lost the luxury of putting my own
needs first (which was what I first thought in the despair of that
moment), but I had lost something almost as significant: the
luxury of being able to choose *when* I could put my needs first.
As time went by, it would become increasingly obvious that the
requirements of my illness would frequently conflict with the
requirements of my children, and that there would be times
when my children's needs would have to come first.

But having children was my choice, you say, and nobody
made me do it. This is true. David and I thought it through,
and looked up the statistics: if one parent has a mood disorder,
a child is roughly three times more likely to develop one too.
But having a child is playing dice with the cosmos, anyway.
Healthy children drown in swimming pools; cerebral palsy

strikes. Besides, if depression did afflict any child of ours (and somehow, in my mind, this would be in his or her teenage years; the possibility of childhood depression never occurred to me), he or she would be much better off with us as parents than with parents who knew nothing about it. The more we thought about it, the more confident we felt. This was manageable if it happened at all, we told each other, and maybe it wouldn't. It was not until much later that we realized we'd applied our artfully constructed logic to only half of the problem. There might be a higher chance that our child would develop a mood disorder, but—barring some miracle cure—there was a 100 percent chance that I would be a mother with one. Years before we would be able to get the first inkling of whether any child of ours showed signs of developing a mood disorder, that child would have to live with the fallout of mine. We were like smokers calculating the risk that our offspring might one day take up cigarettes, never factoring in the effects of years spent breathing polluted household air.

This aspect of the problem became clear with the onset of PPD, of course, but it didn't stop there. When I became pregnant with Rebecca, I immediately went off my medications. After she was born I stayed off—the victim of a delusion common to people with mental illness, the idea that some major life change will magically transform your body, make all things well and whole and new. I was still trying to make a go of it when she was nine months old. I was alone at home one day, Rebecca was at a family day care home down the street, and I was working on a freelance magazine assignment when I encountered a computer problem. Instead of reading the computer manual, I kept tinkering, trying one thing after another. I got irritated. Then I got really angry. Then my heart was pounding.

Soon I was punching the keys instead of typing, swearing, and then I was out of control altogether—screaming, pulling out a chunk of my own hair, kicking the wall, hurling things around my study. This is part of depression, too, these anger attacks, well known in the medical literature if not to the lay public. I hadn't had one in years, but this was the real thing, in full flower, and as always a part of me stood aside, appalled and frightened. I rushed through the empty house, running into doorways, bouncing off walls, and rushed outside, where I found one of Rebecca's toys and kicked it all over the yard. God knows what the neighbors saw. I was sweating all over, breathing hard, and still the rage kept boiling up. I blundered back into the house and went into the kitchen and stood there for a moment, almost reeling. For an instant I thought about getting a knife and cutting myself—anything, *anything* to stanch this volcano in my head—but no: I hadn't done that since I was a teenager, and something sane stopped me now. Instead I turned and blindly kicked the kitchen cabinet. I kicked it viciously, over and over, until the sound of splintering wood stopped me, and I looked. It was a double door, and I had demolished the center support. It was gone.

The anger attack was over, just like that. I remember I sank to the floor and sobbed for a while. Then I drank a stiff shot of Jack Daniel's, even though it wasn't even lunchtime. Then I spent an hour frantically trying to glue the cabinet back into some semblance of what it had been before I had to leave to pick Rebecca up.

David is going to be so mad, I thought. *Now he's finally going to realize how insane I really am. He's going to take Re-*

becca away because I am insane, and I wouldn't blame him.
And I thought, *Thank God Rebecca didn't see this.*

The next day, I called my psychiatrist and went back on my medications.

Leave me alone, my mother always said during her bad times.

The craving for isolation lies at the heart of the experience of depression. Isolation is a prison cell, but at least there you can lie down under your crushing burden. This is not, by the way, a need that can be satisfied by pouring some scented bath beads into the tub and lighting a candle, or going for a refreshing walk, or any of those cheery suggestions you find in the women's magazines. You don't *want* to take a walk with your husband—that's the point: you want to be alone.

But solitude is for writers, artists, and welfare recipients, who are allowed to go into funks. Even people who draw a paycheck at least can claim some sick leave, or "retire in place" at the office, at least until somebody notices. Motherhood is simply incompatible with solitude. You may crave it, but what you get is another "Mommeee!" scraping the surface of your brain like a rusty razor. Unlike ordinary depression, maternal depression is a war fought on two fronts: against the depression itself, and against children's competing demands for nurture, guidance, and reassurance. The barbed-wire boundary between self and others must be painfully navigated every day, several times a day. It takes enormous energy, and that struggle was evident in comments from many mothers I heard from.

"I try to power through the day even though I feel lousy," one mother wrote. Many adopted a kind of lash-yourself-to-the-mast approach: "I grit my teeth (literally)," wrote another. Another mother described getting through her day as "like wading through water with cement shoes." Another mother wrote, "I wanted to curl up in bed and never come out, but the kids needed to eat, to be bathed, to go to their classes. . . . They were completely dependent on me for everything and I could not shirk that responsibility."

"Just the relentless nature of parenting is a monumental effort when you are depressed," another mother wrote. "Putting up the Christmas tree might as well be climbing Mount Everest. The laundry, the dishes, shopping, cooking, homework, stories, bathtime. 'Play with me, Mommy' become words you dread—and then the guilt!"

Oh, yes, the guilt. If by some miracle my children do leave me alone for a while, their unspoken needs pull me out of my isolation, as if an unseen schoolmarm had me by the ear. Children have their needs—and when they sense that what they need may be in short supply, they need it even more. In bad times, I can't even attempt to sleep late without Rebecca appearing at my bedside to announce that she has made me toast, or a concerned Suzanne clambering on top of me to question, "Mommy, my mommy, you sick?"

Even moderate depression can have a direct effect on children. Lucy McHugh, a forty-year-old mother of two who lives in Lynbrook, New York, suffered from a low-grade depression whose main symptom was lethargy; she was shocked when her doctor diagnosed her fatigue as a symptom of depression. But antidepressants made a profound difference in how she felt. Before that, she wrote, "I was not a very good parent to my older child, Jessica. . . . I expected a great deal from her. I had no energy or patience. I found myself yelling at her most of the day for many things that she did not deserve being yelled at for."

In 2002, Marianne Fiedler, of Asbury, Iowa, had a nine-year-old son who suffered from migraines and persistent stomachaches, with no apparent cause. At the same time, she was diagnosed with depression. After several months of treatment, she felt much better. "I [also] noticed Jake had fewer and fewer headaches, and his stomach trouble had pretty much disappeared. . . . Until that 'aha!' moment, I hadn't realized what a profound effect mothers' moods can have on their children."

Having children also sentences you to life. Before, when it was just me during my bad times, I could think of suicide as a reasonable option—not something I wanted to do, but as a trapdoor through which I could escape if life became utterly intolerable. Now that trapdoor is sealed off. Nothing short of getting run over by a Mack truck, or the functional equivalent, will take me from my children before my time. (This sobering realization came to me one day during a particularly bad period as I was looking out our back door at our cat, Ralph, who was torturing a chipmunk on our deck. Ralph was my depression, I thought, and I was that chipmunk, and it was nothing personal.) Imagine, then, craving isolation, feeling ashamed of that craving, loathing myself for setting such a poor example, worrying about letting my daughters see my self-loathing, feeling guilty about not being with them . . . there are subbasements and underground passages of psychological distress here, where once the elevator stopped at the ground floor. Once, when Suzanne and I were alone in the house and I was weeping, she crawled up in my lap and wiped my tears. She was two.

It's easy, during times like these, to flog myself, to say, *You screwup*. But, as my husband frequently has to remind me, I never asked to be sick. And, as I periodically have to remind myself, self-loathing is an extravagant waste of energy. Directness is what's called for. But how? And when? No one wants to burden her children, or rock the foundations of a small person's world; on the other hand, what children imagine can be far worse than what actually is. *If* they imagine, that is; children have also been known to be remarkably blasé about extraordinary things. David and I agreed that we would talk to Rebecca about my illness . . . at some point. Our thinking was no more specific than that. Rebecca was around five, and Suzanne was just a baby, when it became apparent to David and to me that Rebecca seemed to be taking on an inappropriate responsibility for my mood swings, as if they were her fault. So one day when the occasion arose—I don't remember the context—we told her that some people had sicknesses in their brains, and that Mom had one too, and that it caused her to be very sad sometimes. We told her that this sickness had nothing to do with her, and that I was basically okay, but that there might be times when I wasn't feeling myself, and that I would let her know when that happened. But no matter what, we said, things would be fine.

She said very little, and even seemed a little relieved that she was not the cause of the trouble. I congratulated myself on our expert handling of such a delicate issue. But later, David told me that she had asked him the one question every child wonders: "Who is going to take care of me?" When I heard this, I felt my heart crack.

On the other hand, this is a child who has demonstrated

amazing maturity and presence of mind, too, and I do not think this is unconnected. Once, when I'd left her alone with my mother for a few minutes, my mother got her wheelchair wedged behind the bedroom door trying to get to the bathroom. I had stepped outside for a moment, and the only way I could get back in to help was for the wheelchair to move. So Rebecca, who was then six, grabbed that wheelchair by the arm and dragged it and her 175-pound grandmother sideways across the carpet far enough for me to squeeze through. To this day, I don't know how she did it. She is just a child, but she understands a lot. Not long ago, during a period when stress had made me depressed and irritable, I sat down beside her as she watched television. "Bec, I'm sorry about this week," I said. "I've been having some trouble—"

She looked at me. "It's okay, Mom," she said. "I know."

The fact is, my children have miraculous powers, if I can meet them halfway. Sometimes all it takes is getting out of bed. When once I might have stayed under the covers, debating endlessly about whether this particular day was worth living, now there's no time. Even if I feel like shit, I have to make a peanut-butter-and-jelly sandwich for Rebecca's lunch; I still have to get on my hands and knees and locate the long-sleeved pink shirt that goes with her purple pants. I still have to take my pills, and that means I have to eat. Suzanne still wants me to read her a book. And then we might watch *Mr. Rogers' Neighborhood* and then I feel enough energy to tackle the breakfast mess, and so inertia is overcome; the morning passes. Housework is always a useful distraction, and I could easily bury myself in it, especially when I am in a crash—but Suzanne can overcome even my obsessive-compulsive tendencies. "GO! OUT! SIDE!" she yells, tugging on the doorknob, and I think, *Why not?* and get our coats. We walk around the block slowly, taking in the sights—picking up sticks, stomping in puddles, sitting down in the park next to the tennis courts for a few minutes. The journey takes half an hour—it probably seems like a five-mile hike to her—and when we come home we are tired, ready to eat lunch and have a nap. And for an hour or so, I realize, I have felt almost normal.

In a support group meeting not long after Rebecca was born, the conversation one week centered on how determined we all were not to repeat the mistakes our own parents had made. The mood was earnest and misty-eyed. "No, we won't repeat our parents' mistakes," I had said. "We'll make new and even *more* horrible mistakes," and everybody laughed. But it's true. I am avoiding some of my mother's mistakes, but God knows what new ones I'm making. There are no easy answers; the only comfort is in knowing that children are resilient and more generous than adults ever give them credit for. Parenting is hard and human beings fallible; people with no inherent obstacles to success simply abdicate the job every day. Even Anne Sexton's daughter, who had more reason than most to despise the mothering she got, forgave her mother and wrote of her with compassion, pride, and understanding.

If anything, depression keeps me humble. It strips me of any delusions that I am better at this job than I really am. And in the struggle to take care of myself, to navigate that barbed-wire boundary between myself and others so that I can be the best person I can be, I find the courage to be imperfect.

CHAPTER THREE

What's *Wrong* with Me?: When Maternal Depression Starts with a Bundle of Joy

I was afraid of her and ashamed of myself. I used to watch [my baby] sleeping and wonder with pride at how beautiful and perfect she looked. Then I would cry because I felt so sorry for her for her having such a screwup for a mother.
—KATHERINE SCRIVER, TWENTY-SIX,
OF FORT WAYNE, INDIANA, MOTHER OF NORA, TWO

On May 10, 1876, thirty-year-old Elizabeth Dreher was admitted to St. Elizabeth's Hospital, a charity hospital for the insane in Washington, D.C., with a diagnosis of "acute melancholia due to childbirth"—what we would most likely call postpartum depression (PPD). Her four-week-old son was taken from her and placed in the Washington Orphan's Home. The whereabouts of her husband are not mentioned in her file, nor is the existence of any other children. Five months later, one R. Gallause—perhaps an attorney hired by

Mrs. Dreher's friends, perhaps just a friend—wrote to the superintendent of St. Elizabeth's, C. H. Nichols, asking that Mrs. Dreher be released to the custody of a Mrs. Fahrnkopf, a resident of the District, who had offered to take care of her. The letter also asked about the whereabouts of Mrs. Dreher's baby. There is no answer to this request in the file, but the request was evidently unsuccessful: the next entry, dated May 14, 1877, is a letter from the District of Columbia police department, asking Nichols what they should do with Mrs. Dreher's household property.

Sixteen years later, Elizabeth Dreher died of tuberculosis, still an inmate at St. Elizabeth's. The last two items in her file were written in April 1898, six years after her death. The first item is a letter from John Dretcher (as he spelled his last name), the infant Elizabeth Dreher had last seen when he was four weeks old. Adopted at the age of two by a couple in New York, Dretcher was now twenty-one and searching for his mother. The second item, dated a few weeks later, is also a letter from Dretcher, who had evidently been informed of his mother's death. "Did anyone call for the body of Mrs. Dreher?" her son asked. "Did anyone call to see her or look after her all the while she was there?"

We have come a long way since suffering from PPD could wind up in being institutionalized for life. Celebrities talk about PPD these days; Brooke Shields and Marie Osmond have written books about their experiences. But the fact that their stories are considered noteworthy is itself testimony to how little the public understands about this illness.

Many people assume, for example, that maternal depression and postpartum depression are the same. In fact, PPD is just one chapter in the story of maternal depression; many mothers who struggle with depression deliver their children happily and encounter depression only when their children are past infancy. However, having a history of depression prior to childbirth makes PPD more likely, and suffering one episode of PPD puts you at higher risk for having PPD should you have another baby. This is all the more alarming when taken into account with the fact that PPD is all too often left undiagnosed and untreated. Judging from my interviews and the responses in our survey, a mother who suffers PPD today has roughly a fifty-fifty shot of encountering an obstetrician who is significantly better prepared to help her than were the doctors of Elizabeth Dreher's day.

PPD is important to the discussion of maternal depression for two reasons. One is that it can cause disruptions in the mother-child relationship that can have lasting repercussions. The second is that depression is by nature a cyclical illness. Having suffered one episode of PPD increases the likelihood that it will recur after another childbirth, at least for most women; for women with a history of depression, an episode of PPD increases the likelihood of future depressive episodes in general.

There is lingering stigma associated with discussing any kind of depression, but the stigma associated with PPD is especially fierce. These days, it's probably easier for a man to talk about problems with his penis than it is for a new mother to admit that (in her own eyes, anyway) she is a failure at the one task for which women are supposed to have instinctive gifts. "I have never felt so alone and ashamed in my life," wrote Deborah Chaney, an artist from Hermosa Beach, California. "Even now I find it hard to talk to other moms who seem to have it so great."

The trap that awaits mothers who speak frankly about the difficulties presented by motherhood is that anything they say may be—often is—interpreted by listeners as sniveling. Haven't they just had a *baby*? How can there be anything bad about *that*? "I fear people will misunderstand if we whine and complain about the loss of freedom, independence, and privacy, even though we wouldn't trade motherhood for anything," one survey respondent wrote. "I sometimes just want to be heard and validated. Then I can keep going." And women can be terrific at doing that for other women. But the flip side of that capacity for close emotional connection is that women also know just where to shove the knife. "The one friend I tried to discuss it with basically told me to stop being whiny and look at the REAL problems in the world and take care of my babies," wrote another respondent. "I got no help with it and really did not realize what the problem was until years later."

That kind of ignorance still extends to big chunks of the medical profession.

Six weeks after Meredith Goodlatte, a thirty-four-year-old former pharmaceutical rep from Orlando, gave birth to her daughter, Madison, in 2001, she found herself unable to bear the sight of food, subject to uncontrollable crying spells and suffering from mounting anxiety. At first, she said, she thought she was pregnant again—"a scary thought with a six-week-old." Determining that she was not, she made an appointment with her obstetrician, "who said I was fine and these symptoms were normal." A week passed, and the anxi-

ety escalated into full-blown panic attacks. She was so inca-
pacitated that she could barely get out of bed. Her husband
took her back to the obstetrician. Meredith remembers weeping in his office and hearing him say, "I delivered your baby
and now if you think something is wrong with you, get a family doctor."

How many obstetricians out there would display a similar
indifference? There's no way to know precisely, but judging
from the responses to the survey, a safe answer would be "too
many." (Meredith Goodlatte got help via a neighbor, a nurse
practitioner who worked in the office of a more enlightened
obstetrician.) To be fair, many of the mothers went out of
their way to credit their obstetrician with recognizing their
symptoms and intervening quickly. But of the 210 women
who had experienced at least one episode of PPD, an average
of 55 percent said their obstetrician was of no help. Twenty
percent said their obstetrician was "somewhat" helpful. Only
25 percent said their doctor was "very" helpful—hardly a
glowing review for the medical specialty that, more than any
other, is likely to encounter patients with this illness.

It's not as if PPD is a new phenomenon. Its written history
goes back hundreds of years. In 1436, Margery Kempe, the
twenty-year-old wife of a city official in Lynn, England, gave
birth to what would be the first of fourteen children. In her
journal—in which she refers to herself in the third person—
she writes that after her child's birth, "She despaired of her
life." During a period she records with some precision as "half
a year, eight weeks and odd days," Kempe writes that "she
would have killed herself many a time . . . and in witness of
this she bit her own hand so violently that the mark could be

seen for the rest of her life." In 1817, American physician Thomas Ewell's treatise on pregnancy and childbirth noted, "Women after delivery are very subject to headaches and to madness."

Ewell's description may sound patronizing to modern ears, but for any woman who has been through this experience, *madness* is not too strong a term. And confusing PPD with "the baby blues," which many laymen and doctors do, is like equating a summer shower to a hurricane. "I will never have another child because of this," one mother in the survey wrote grimly, describing her experience with PPD. "We made sure of that." Even in its milder forms, PPD can make a woman feel like a stranger to herself—a new being with the IQ of a cow, resident of a reality where all perceptions are subtly distorted. "I felt like I was in a tunnel," wrote Sandra Coleman, of St. Joseph, Illinois, "and things sounded like they were underwater." To Stephanie Walters, who lives in Meridian, Idaho, it was "the strangest thing that has ever happened to me. . . . I walked around like I was a robot, watching myself without any emotion." Memory goes, too; details are foggy, and time distorts—growing impossibly long in some places, erased in others. The first three months of her daughter's life are now a blur to Kathryn Brown, who encountered PPD after Ava's birth in 2003. "I can't remember the smell of her little head after a bath," she wrote from her home in Charleston, West Virginia. "I hate that I lost that time with her." Another survey respondent wrote, "I cannot remember the first two years of my second child. I look at pictures and honestly don't remember anything. I wish I could go back and live those days again."

Sometimes PPD is a continuation of a depressive episode that began during pregnancy. At other times, its onset is like an anvil falling from the sky—a terrifying mental unmooring that begins, researchers believe, with the dramatic drop in estrogen, progesterone, and cortisol levels in the mother's body after birth. That drop is normal, and what's normal also is that estrogen and progesterone play an important role in helping to control the action of the neurotransmitters serotonin, dopamine, and norepinephrin, low levels of which are known to occur in depression. Current research suggests that some women have some genetic difference in the way their brain's HPA axis (see page 13) operates that makes them particularly sensitive not to hormone levels per se, but to the rate at which they drop.

Depending on the exact nature of the biological vulnerability, the result is major depression or, for a smaller subset of women, manic "highs," an anxiety disorder known as postpartum obsessive-compulsive disorder, or, in extremely rare cases, actual psychosis.

But there is a lot we still don't know. For instance, women with certain types of depression have extremely high cortisol levels, and so do women in late pregnancy. Why aren't *all* pregnant women depressed? On the other hand, estrogen levels start to rise at the beginning of pregnancy and, by the end, are roughly four hundred times their prepregnancy level. If a drop in estrogen after childbirth helps to trigger depression, high estrogen levels in pregnancy should protect women against depression, right? Wrong. Recent research has shown that depression is at least as common during pregnancy as it is during any other time of a woman's life, and many cases of

PPD are simply a worsening of depressive symptoms that were present before childbirth.

The cause–effect chain is obviously complex, and adding to the complexity is that biology isn't all there is: environment, as always, also plays a role. Women who are socially isolated are more vulnerable to the disorder, as are women who are in troubled marriages, women whose pregnancy is marked by some major life event (like the death of a parent), and women who had their babies under particularly stressful conditions— after an extremely long labor, say, or an emergency C-section.

Still another factor is an infant's temperament. In a way, every birth is like a blind date: babies come from the womb with their own unique personalities and, as with blind dates, certain personalities fit together better than others. Infants who are irritable, who are difficult to soothe, or who cannot seem to adjust to a schedule can bring out the worst in a mother who is suffering from PPD; she is inept at soothing her infant, who gets even fussier, and a vicious cycle is established. Or a fussy baby can cause severe sleep deprivation, which increases the risk of depression. Shari Lusskin, director of Reproductive Psychiatry at New York University, goes so far as to say that sleep deprivation is "the root of all evil" when it comes to PPD.

It's an evil unwittingly perpetrated by the movement to get more mothers to breast-feed. Breast-feeding is without a doubt the best possible thing to do for anyone who can manage it, but to be entirely successful, it requires the mother's twenty-four-hour-a-day availability for months—a demanding commitment for a woman in the best of health. This pressure can go to excruciating lengths: a friend of mine who was suffering from PPD after the birth of her first child was told by her doctor that to increase her milk supply she should nurse on one side while latching her other breast onto a mechanical breast pump. Exhausted, in pain, and desperate for sleep, she snapped, "Yeah, and if I stick a broom up my ass, maybe I can sweep the floor at the same time." (Her doctor appeared to be baffled by her reaction.) Without sleep, the symptoms of PPD only get worse. Yet pressure from well-meaning friends and doctors to breast-feed ignores the sufferer's primal need to get at least eight hours of uninterrupted sleep at night (while someone else, presumably Dad, handles night feedings with a bottle). The American Academy of Pediatrics' latest policy statement on breast-feeding, issued in February 2005, lists only a few medical reasons not to breast-feed: if the mother is under chemotherapy or radiation therapy, has certain rare diseases, abuses street drugs, has herpes, or is infected with the HIV virus. Preserving a new mother's mental health is not, apparently, a good enough reason to give a baby formula.

Then there are those two dreaded words: "infant colic"—words that, when put together with PPD, are a combination made in hell. Women have an innately greater sensitivity to noise than men, and women who have just given birth are exquisitely attuned to the sound of a baby's cries. (For some breast-feeding mothers, the letdown reflex that starts breast milk flowing can be triggered by the sound of some other woman's baby in another aisle of the grocery store.) Imagine, then, being Dalisa Madaski, whose second baby, born in 2003, "cried almost every waking moment of the day, six to eight hours a day" for the first four months of her life. "I really can't explain to you my mental state at this point [from four] months of listening to a baby cry," she wrote. One day, sitting in the car in the driveway of her home in East Lansing, Michigan, Dalisa finally lost it: "I broke my eyeglasses with my bare hands, punched my dashboard, kicked my bags from the grocery store," she wrote, while her husband, seven-year-old son, and the baby watched in amazement and horror. Dalisa called this "a ridiculous fit." After four months with a colicky baby, it seemed like a fairly restrained reaction to me.

Before that episode, Dalisa wrote, she had not realized she was suffering from PPD. And that is the ultimate paradox of PPD: the worse the symptoms, the harder it is for a woman to recognize in herself.

When Meredith Goodlatte's neighbor showed her a list of thirty possible symptoms of PPD, she found she had been suffering from twenty-seven of the thirty. How can anyone be that sick and not know it? It's easy. Other illnesses deceive, too: high-altitude sickness, for example, or heat stroke. And for first-time mothers in particular, PPD comes at a time when life has been transformed and everything is in flux, so it's tempting to conclude that disorientation is normal. Another reason is that PPD can begin very differently from the way a woman has experienced depression up to then—with anxiety and insomnia, for instance, instead of emotional numbness and feeling physically and mentally "slowed down." To make things even more confusing, symptoms in the same woman can vary from one episode to the next. Stephanie Walters, who had PPD after the births of each of her two children, suffered debilitating depression with the first, intense anxiety and irritability with the second.

Even women who manage to find appropriate medical treatment often do so only after enduring months of unnecessary suffering. In the general population, roughly 10 percent of women who experience PPD still show signs of it after one year. Of the women in the survey who had PPD, 52 percent waited at least six months to seek medical help, and most of those waited more than a year. Other women wind up going to counselors who shoehorn their symptoms into some outdated diagnosis, as did Jeannie Wegman, of Cedar Lake, Indiana, who saw a counselor when she suffered from PPD after the premature birth of her first child in 1996. The counselor told her that her symptoms were a manifestation of her "unconscious resentment" toward her baby.

"Did you have postpartum depression after we were born?" I asked my mother. We were sitting on a porch outside the nursing home in middle Georgia where she lives. This was maybe five years ago; Rebecca was a preschooler, and Suzanne had not been born. It was spring, which in middle Georgia means at least 80 degrees. I remember that day as sticky and hot; sweat was pouring down my back. My mother was wrapped in a sweater, barely keeping warm.

This was a question I'd never asked her before. Long ago, I'd learned to steer away from queries about her emotional life. She had always been tight-lipped about such things, in the way people often are when their childhoods were marked by deprivation and abuse. So it was strange that I asked her, and even stranger that she answered readily. "Oh, yeah," she said. "But we didn't call it that then."

"Was it worse after I was born, or after Nonny was born?" Nonny is my sister, born two years before me.

"Oh, with Nonny, I think."

"Did you ask for help?"

"Oh, no. That's just the way it was."

"Did you talk to Daddy?"

"No. He didn't understand." She paused, and I began to think her mind had wandered—which, since her stroke, it tends to do. Then she said, "But you were good babies."

"What do you mean?"

"I could put you down on the floor beside my chair with a coloring book or something and say, 'Mommy needs to rest for a while,' and you would just play there quietly, and I could put my hand over my eyes and cry so you couldn't see."

It is that red vinyl recliner I imagine her in. In this made-up memory—because in fact I don't consciously remember this at all—it is late winter, and the house is neat and quiet. Maybe my sister is in school. There is a gas heater in the room, on low, so the only sound is a low, companionable hissing noise. I am the child coloring on the floor beside the woman who is weeping, who thinks nobody knows she is weeping.

Rebecca was born in late November, on a Thursday night. Her birth, which came two weeks early, was the end result of a series of events that had begun two days earlier, when the nurse at my office discovered that my blood pressure had spiked to something like 150/110. In quick order, I found myself in the hospital, hooked up to a fetal monitor. I wanted to avoid a caesarean if possible, so my doctors decided to induce labor that night with a prostaglandin suppository. Early the next morning, they hooked me up to a Pitocin drip, and what had been manageable discomfort suddenly turned into the feeling of being run over by a freight train at three- to four-minute intervals. It was 3 P.M. before I was dilated enough to get an epidural. At 6 P.M., I awoke from an exhausted nap to find out that my labor had stalled, and I was going to get a caesarean after all. That was the easy part.

We went home on Monday. At 5 A.M. on Tuesday morning, I was sitting in bed, drenched in night sweat and trying to get a screaming baby to latch on and nurse, when—and this is the only way I can describe it—a howling blackness descended on my mind. Time stopped. My life was gone; the moment telescoped suddenly into an eternity where I was tethered to a baby whose needs I could not meet, would never be able to meet, and who was probably dying for lack of my milk. The world outside the bedroom window did not exist. I remember that the book that happened to be on my bedside table featured on its cover a smiling woman holding an infant; its title was *Relax and Enjoy Your Baby!*—and, as if the title weren't bad enough, that cheery exclamation mark at the end was the final, mocking blow. Despair hit me like a poleax. I turned to David. "This is hell," I told him, "and I am in it, and I am going to be here for a while."

The weeks that followed live in my memory as one long nuclear winter. Every time she cried, it felt like someone had taken an electric cattle prod to my spine. A white bolt of nervous energy ripped from head to foot and tore at my insides: the first few nights, every time I heard that hungry wail, I had to choose between going to her or running to the bathroom with diarrhea. I was so wired I could not even lie down for long, much less get any restful sleep. Breast-feeding was a disaster; it's hard to nurse a baby with anxiety as intense as mine. Rebecca would latch on, find no milk to speak of, and start screaming in frustration, and we repeated this cycle until my nipples bled. Day and night became one long twilight. My sense of time warped, bent, and then went entirely. Four hours could seem like five minutes; five minutes could seem like all day.

At the same time, my illness marched to a strict diurnal rhythm: every afternoon at four, as the light began to fade, I would begin to cry uncontrollably. "I'm drowning," I told David, writhing with sobs facedown on our bed. David curled up behind me, cradling my body with his. "We will get through this," he said. "We will get through this." I don't know if he truly believed it at the time, but his reassurance was the only thing I had to hang on to. My mother was ill; my sister could not get time off from work; my husband's mother had died when he was fourteen. I was surrounded by men who loved me and who performed heroic feats to help me—my husband, my father-in-law, my brother-in-law—but somehow that added to my sense of isolation: I craved the company of another woman. I remember at one point sitting on the living-room sofa, telling David I needed to go to the hospital, and realizing that this frightened him—which told me that not even he fully appreciated how much trouble I was in. Seeing his fear, I backed down. And yet, had anyone asked me at the time, I am not sure I would have said, "I am suffering from postpartum depression." I would have said I was suffering from entirely understandable grief at my own incompetence.

At my six-week checkup, I hesitantly mentioned to my doctor that I had been feeling, well, a little sad. My doctor was a genial older man who had delivered thousands of babies. He smiled reassuringly. "I had another patient who said that," he said. "I told her, 'You can't be depressed! You have a baby to take care of!' " My first thought was that from my position sitting on the end of the examining table, I was in an excellent position to give that man a swift kick in the balls. But that lasted only an instant; immediately, I felt ashamed. Depression has its own ironclad logic. *Either all moms go through this without complaining, so I should just shut up,* I thought. *And if they don't, the fact that I'm miserable means I'm a lousy mother.*

For the next year, my depression was like a shark. At times it would drag me under and chew on me; at other times, it would let me go for a while, pondering its next move, while I floundered to the surface. But it was never far away, and it seemed to me there was no one to help. Every time I talked to David, it added to his burden and cast a pall on his happiness; after a while, his feeling of helplessness began to morph into anger—not at me, but at the situation. As for my psychiatrist, I was loath to tell him how badly I was failing at my new job. I was afraid that he would want me to go to the hospital or, at the very least, insist that I go back on my medications, which I was determined not to do because it would interfere with breast-feeding. I had never given the subject of breast-feeding much thought, except to assume that it was something any old mammal could do. Now, it represented the one last thing I had not yet screwed up, and I was damned if I was going to fail. But my body would not cooperate, not even with the help of a lactation consultant and an industrial-strength breast pump. After about two months, I finally gave up. That did it: I was now, officially, a Bad Mom.

What saved me, I think, was that one day in January, I had a brief moment of clarity: *You need to talk to other women*. I found a new mothers' support group, and they became my life raft. I hung on to them and white-knuckled it until, late that summer, that episode with the kitchen cabinet door forced me to face the fact that I needed my pills.

It was a rough introduction to motherhood. But, as I was later to discover, it wasn't that unusual.

Laurel Spence is a tall woman with neatly coiffed blond hair, an athletic build, and the kind of gangly gracefulness Southern girls acquire when they discover they will never fit the sweet petite Southern ideal. She is thirty-three, a physician's assistant in the office of Dr. Paul Cook, a prominent Houston obstetrician. She got pregnant with her first child in 2000, when she was finishing up the two-year physician's assistant program at Baylor University and her husband was working in computers at the University of Texas. Madeleine's birth was uncomplicated. But soon after, Laurel began to suffer unusual anxiety.

It began with breast-feeding. "I was a total perfectionist," she laughed. "Every one and a half hours, that baby had to be put to the breast." She became obsessed with schedules. Visits from friends became unwanted intrusions. The ringing of the telephone sounded like a fire alarm. She began losing weight at an alarming rate. Insomnia set in. She couldn't concentrate on what she read, and if she did read something she couldn't remember what it was. As a medical professional, she caught on to the problem fairly quickly: "It was pretty much across-the-board symptoms of [postpartum] depression," she said. But she told herself: no pills. Somehow, she toughed it out for a year, and then the depression wore off.

In July 2002, she gave birth to her second daughter, Kate. By then, she had left her first job, which she had found stressful, and had joined Cook's practice. She loved working there, and especially liked working with Cook. "Have you ever seen *Dr. T and the Women*—that movie where Richard Gere plays an obstetrician?" she said. "Dr. Cook is sort of like that. To his patients, he's sort of a Greek god or something."

She told me this one day in the summer of 2004 as we talked in her office, in a downtown Houston skyscraper. The room was tastefully appointed, with gray walls decorated with Rothko-esque abstract art and anatomical models of the female reproductive system. Outside, a typical Texas summer thunderstorm was emptying tons of warm rainwater onto steaming asphalt.

"So when Kate was born, that was when you got medical help, right?" I asked.

Wrong. She told no one. "I was afraid," she said. Of what? Of that mocking voice in her head, the one that kept saying, "You are such a wimp"—the same kind of voice I had heard after Rebecca's birth. At Laurel's six-week postpartum checkup, she didn't mention her depression, and her doctor, one of the doctors in Cook's practice, didn't ask. Soon she had lost her baby weight, and the pounds just kept dropping off. "I wasn't making meals, and even if somebody did make one for me I wouldn't pick it up," she said. Every day became a little bit harder than the day before, as if she were trudging up a hill that kept getting steeper. There she was, at a job she loved, a sturdy marriage and two beautiful little girls, and she could not escape an oppressive sense of being a failure, of having too much to do and not doing any of it well. Like many Southerners, Laurel grew up in a devoutly religious environment in which she was taught that there was no problem too big for God. Yet this problem seemed to be, because things weren't getting better. She added "lack of faith" to her list of things to feel guilty about. It was getting to be a long list.

After three months, she decided to avail herself of the office supply of Zoloft samples, and secretly started herself on a daily dose of 50 milligrams. It helped, a little. In July 2003, Laurel decided to celebrate Kate's first birthday by taking herself off the Zoloft. It had been nine months, she thought. Surely the depression had run its course. Wrong again. The old symptoms returned in force: anxiety, inability to concentrate, insomnia, loss of appetite, hyperirritability. "I was a nut case," she said. "Mostly, I felt really, really overwhelmed. And guilty, for working and being away from the kids." It was only then, after nearly a year of misery, that the wall of denial and stoicism crumbled. She made an appointment with a therapist, who took one look at her and immediately doubled her dose of Zoloft to 100 milligrams a day. Within weeks, she began to feel better.

Today, Laurel is fully recovered, though she is still on a maintenance dose of Zoloft. Coming out of the closet about her PPD has changed not only her life, but the way her employer practices medicine. Paul Cook is an experienced and respected obstetrician, but working alongside a fellow medical professional who felt ashamed of being sick opened his eyes to the possibility that just maybe he was missing something important in his patients as well. Before Laurel's illness, Cook told me, "we weren't doing much at all" to screen new mothers for PPD—even though the infamous case of Andrea Yates, who killed her five children while in the grip of postpartum psychosis, had happened in a Houston suburb just a few miles away. The Texas courts dealt with Andrea Yates by sentencing her to life in prison, and the Texas legislature responded by passing a law requiring obstetricians to document that they ask their postpartum patients whether they are suffering any symptoms of depression. Cook regards the law as another piece of paperwork. How is one perfunctory question going to turn up anything when his own assistant couldn't bring herself to admit it for months?

It was clear to me that Laurel was still struggling with a residual need to apologize for her illness. "My husband married me because he thought I was so stable!" she blurted out during our interview, and laughed self-consciously. Later, in a note she sent after our meeting, she described a kind of waking dream she'd had the night before—an image of herself as Hester Prynne. But instead of a scarlet A on her chest, she wore a scarlet D. "I can choose to hide behind the scarlet letter . . . or I can choose to unmask it, learn from it, [and] move forward," she wrote.

She has chosen the latter. Today, any woman who displays symptoms of PPD is referred to her for counseling and the kind of support that can only come from someone who has lived through the same thing. Laurel has also created a package of information that is now routinely handed out to all pregnant women as they near their delivery date. The packet contains a list of symptoms to be aware of and a list of local and online resources. At their six-week postpartum checkup, women patients are now given something called the Edinburgh Postnatal Depression Scale—a five-minute questionnaire that is a standard instrument used to screen for symptoms of PPD. That brings Cook's practice up to the standard of care that the American College of Obstetricians and Gynecologists has been recommending since at least 2002. It's a standard much higher than that required under Texas law, but it doesn't have the force of law—and judging from what women in our survey said, many obstetricians simply ignore it.

Why the huge blind spot? To begin with, obstetrics is a field that attracts people who want to deal with healthy patients expecting happy outcomes. They may be temperamentally unsuited for detecting signs of illness, especially psychiatric illness. Another problem is that PPD falls on the line dividing two specialties—obstetrics and psychiatry. Finally, in a specialty as litigation-prone as obstetrics, it's easy for doctors to unconsciously begin practicing defensive medicine: don't ask questions you don't want to know the answers to. It's not hard to imagine a great many scenes like the one below, described by one of our survey respondents:

"My ob-gyn asked if I [had] 'any other problems' at my follow-up visit after the delivery. I asked if he meant depression and he uncomfortably said, 'That is a possibility.' I responded that I had been feeling depressed. He asked if there was anything he could do. Since I couldn't understand how a gynecological exam would assist me [and] I didn't think he could/would prescribe an antidepressant, I said, 'I'll be fine.' He turned and left the room. I now wonder how many women he has [had] this uncomfortable conversation with."

Cook, who has been in practice for thirteen years, doesn't recall much if any mention of PPD when he was a medical student, which would not come as a surprise to Dr. Kirsti Dyer, a Sonora, California, mother and physician. She has seen the issue from both sides: she doesn't remember anyone asking about how she was doing after her own two children were born, even after one of them spent several weeks in a neonatal intensive care unit. As for her colleagues, she wrote, "The mother has to be pretty bad before someone picks [PPD] up. It is not something we [were] trained to ask about."

Both Cook and Dyer are doctors who have spent most of their years of practice in the post-Prozac era; a doctor who entered practice before then is even less likely to notice symptoms.

Finally, medicine is an inherently conservative profession. The first edition of Dr. Benjamin Spock's *Baby and Child Care,* published in 1945, included a section on "the blue feeling," as Dr. Spock called it. "Try to get some relief from the constant care of the baby in the first month or two, especially if he cries a great deal. Go to a movie, or to the beauty parlor, or get yourself a new hat or dress," he advised, and then added, with cheerful illogic, "Take the baby along if you can't find anyone to stay with him." The fiftieth anniversary edition of his work, published in the mid-1990s, does use the term *postpartum depression.* Otherwise, it reads word for word exactly as it did in 1945.

Postpartum obsessive-compulsive disorder (OCD) is a relatively rare subtype of PPD in which the mother is tormented by intrusive thoughts about harming her baby. It is as if her maternal instincts have gone into overdrive: not only does she see potential danger everywhere, she can all too vividly imagine it. Like PPD, it is linked to plunging levels of specific hormones—cortisol, progesterone, and estrogen—and the effect of those changes on neurotransmitters in the brain. It tends to occur in women with a history of anxiety disorders or a family history of OCD.

Sunshine Gage teaches psychology at a community college in the Houston area. She is a small woman with streaked blond hair and a wide-open smile who lives up to her name, although there is a touch of wariness in her manner when she met me, as if she were halfway expecting unpleasant news. She is married to a roofing contractor. Their son, Ace, is three.

When she became pregnant with Ace, she was working at a research center affiliated with Texas Medical Center, specializing in psychiatric research, but postpartum OCD was a topic she'd never had any reason to look into. In retrospect, she thinks she was primed for it—both by genetics and by a sometimes chaotic childhood. Her family is close, she said, "but when we fight, it can get really ugly . . . things being thrown, broken and smashed, yelling and screaming, out-of-control anger." For a child, "it [was] definitely traumatic." Perhaps as a consequence of that, she has always been anxiety-prone, a child who would stay up until 3 A.M. worrying about burglars.

In her sixth month of pregnancy, she began having obses-sive thoughts about her brother, who was at the time in a tu-multuous romantic relationship. "It was a visual obsession," she said, in which a vivid image of her brother getting shot played in her mind over and over. The obsession lasted about a week, and then faded. It was an omen.

Ace's birth was uneventful. Then, when he was two months old, Sunshine was sitting on her living-room floor with him one morning, watching him with that awestruck wonder new mothers feel. "My baby was lying there, and I was just admiring him," she said. She thought of all the sto-ries she had heard about child abuse, and "I was wondering how could anybody hurt something so precious." At that in-stant, an image popped into her mind: a man bent over Ace, sexually molesting him. She could not get the picture out of her mind. She was horrified at herself. Who would even *think* of something like that? What kind of *mother* could dream up such a thing about her own baby?

That night, she told her husband what had happened. He was not alarmed. "I've always had a confessing compulsion," she said; she was known for telling her boss about the tiniest mistakes she had made, or telling her husband about almost every thought that crossed her mind. To her husband, this was just another example of her little quirk. When she men-tioned her disturbing thoughts to a few close friends, they also shrugged it off. "People would say, 'These are just crazy thoughts; you are a wonderful mother.'"

The intrusive images in her mind changed from the sexual to the graphically violent, always involving her baby. She coped by inventing tasks and errands, keeping herself so busy that she had no time to think. But secretly, she said, "I felt like the worst person in the world." She was another Andrea Yates, she thought; she was worse than a common pedophile. She called her obstetrician's office, where the nurse practitioner told her, "What you are feeling is normal." The only thing she was sure of was that it was *not* normal. Next, she called her family doctor, who prescribed Paxil and warned her that it was not safe to take while she was breast-feeding. She quit breast-feeding, and tried Paxil. It didn't help. So she tried Klonopin, an antianxiety drug. "It didn't help the intrusive thoughts," she said. "It just took the edge off." Next she tried Prozac, which didn't help either.

The sickening, violent mental images were more intrusive when she was alone, and when she was alone with Ace they were inescapable. Night and day, she was haunted; it was as if some sadist were forcing her to watch a sick snuff movie, and she could not turn off the projector. "It got so bad that I thought, 'If someone thinks I'm capable of this, it would be better if they put me away.' "

Two months passed. Desperate, she sought out a psychologist. "I think you're depressed," the therapist said, and they began talk therapy. And then—

As we had been talking, Sunshine had been nervously pushing her hair away from her face, and now she gathered it up in both hands behind her head. "This is what made me an activist," she said. "This therapist said, 'I know what's wrong with you—I saw a segment about it last night on *20/20*.'" Finally, the correct diagnosis—from a TV show.

Now that she knew what to look up, Sunshine used her research skills to find everything she could about postpartum

OCD. Along the way, she ran across the name of Lucy Puryear, a Houston psychiatrist and specialist in postpartum disorders (and an expert witness called by the defense in the Andrea Yates case). Dr. Puryear put her on a low dose of Zoloft, reinstituted the Klonopin for anxiety, and then added Seroquel, an antipsychotic drug. She also recommended cognitive behavioral therapy, a type of therapy based on the concept that, just as emotions create thoughts, thoughts can alter the neural architecture of emotion. Cognitive therapy emphasizes symptom control and mental "homework" designed to create new habits of thinking. The combination worked. Ten months after that day on her living-room floor, she was back to feeling normal.

If the stigma surrounding PPD is considerable, it is magnified tenfold when it comes to postpartum OCD. How can a mother summon up the resolve to say the words, "I can't stop thinking about dropping my baby off the balcony"? What if she is desperate and nobody takes her seriously? Even worse, what if they do?

"I could not sit in the kitchen with a sharp knife on the countertop," wrote a thirty-seven-year-old pharmacist who lives in Laurel, Maryland, who experienced postpartum OCD when her twins were born in 2002. "I was afraid I would hurt them in the car while my husband drove and he could not stop me. I had nightmares that I would do something when I got up to feed them during the night while everyone else in the house slept. . . . I thought, Whoever finds out about this will surely take my babies away from me. I must be crazy."

In fact, up until the early 1990s, mothers who confessed to such obsessions were classified as "homicidal mothers"—a term that conflates postpartum OCD with another condition entirely, postpartum psychosis. Women with postpartum psychosis (which is extremely rare) usually operate under a delusion they firmly believe in, and are often unable to tell right from wrong. A woman with postpartum OCD is tormented by the idea that she might do something wrong; a woman with postpartum psychosis may simply act on her delusions, believing she is doing something difficult but morally right. (Andrea Yates, for example, testified that she killed her children in the belief that she was saving them from Satan.)

Yet these distinctions still elude many doctors. Sunshine Gage now has a second job of sorts, as an activist in the field of educating women about postpartum disorders in general and postpartum OCD in particular. She sits on the board of a nonprofit group called the Postpartum Resource Center of Texas and, as word gets out through the Internet grapevine, has found herself in touch with women all over the country who have suffered similar experiences. Toward the end of our interview, Sunshine told me the story of a friend who also suffered postpartum OCD. When her friend told her obstetrician about it, she said, "he cut her off and said, 'Oh, that's psychosis,' " she said—in effect, presenting his patient with the choice of renouncing her symptoms or getting reported to social services. Sunshine paused, and gave a short laugh. "And [doctors] wonder where we get our motivation."

Suzanne's birth came four years and five weeks after Rebecca's, and by then I had learned a lot. I found a sympathetic and informed woman obstetrician, and when I found out I was pregnant, I tapered off the Effexor I had been taking and started on Prozac. Suzanne's birth was a planned C-section, and except for an epidural that never did completely "take," her birth was joyously uneventful.

Even so, on day three after her birth, I was jolted awake from a nap in my hospital bed. My heart was pounding. It was the same white-hot anxiety I had felt after Rebecca's birth, as familiar to me as if that had been yesterday. My father-in-law was in the room, and I was too ashamed to say anything in front of him. Instead, I waited until he left, and then I forced myself to tell my husband what was happening. "Call Dr. Diamond," he said. "Right now."

Michael Diamond, my psychiatrist, prescribed a low dose of an antianxiety drug and switched me from Prozac to Paxil, at a higher dose. The next few weeks were uneventful. The anxiety subsided. We hired a doula to help with the baby, and I got as much sleep as possible. This time, I did not try to be a hero about breast-feeding. Without around-the-clock feeding, I knew it would end early, and it did. Meanwhile, I loved seeing Suzanne smacking her lips, looking drunk with milk, or getting ready for a meal with a merry look in her eyes, as if she were thinking, "Ohboyohboyohboy." Once in a while she would stop and gaze at me, as if to say, "Let's just savor this moment, shall we, Mom?" At other times, when I interrupted her to switch her from one side to another, she would get this wild "where is my next meal coming from?" look on her face. It was hilarious. *This,* I thought, *is why women love having babies.*

Everything was different, and easier, now that I knew what I was dealing with. And, for whatever reason, Suzanne was a much happier baby than Rebecca had been—a cuddly girl who quickly settled into a routine. Ignorance had been my enemy when I had Rebecca. Getting a second chance taught me that with the right expectations and the right medical advice, my baby and I could do just fine.

But if ever there was a subject that requires being a full partner in your medical care, it's this one. Knowledgeable and sensitive obstetricians are out there, but you cannot simply assume (as I did in choosing the doctor who delivered Rebecca) that years of experience equate to first-rate care. Take, for example, the two most common questions that come up when a woman is dealing with depression in the context of pregnancy: Is it safe to take antidepressants during pregnancy? and, Can I take antidepressants while I'm breast-feeding?

Many of the women who responded to the survey told of doctors who advised them to go off their medications during pregnancy or told them there was no such thing as an antidepressant that could be taken while they were breast-feeding. It's advice that is correct about some antidepressants, and physicians who give it are following the conservative medical adage: when in doubt, do nothing. "Safety" is also a relative term. Aspirin is perfectly safe—unless you are a hemophiliac, in which case it may make you bleed to death. There are no double-blind, placebo-controlled studies about the use of antidepressants during pregnancy, because such studies would require both withholding treatment from women who are ill and subjecting fetuses and infants to treatments they cannot give their consent to. But drug companies keep track of case histories of women for whom the risk-benefit calculus tilted toward continuing medication, and over the years this database has yielded scant evidence of any adverse effects on fetuses or breast-feeding babies from Zoloft, Paxil, or Luvox.

What's more, emerging research suggests that women with a history of depression who go off their medications during pregnancy will usually relapse within a few months. The decision to discontinue medications should involve weighing the known severity of the woman's illness versus the hypothetical risk of harm to the fetus, said Dr. Lori Altschuler, director of the Mood Disorders Research Program at the University of California, Los Angeles. Potential risks to the fetus from drug exposure, moreover, should be weighed against a formidable body of data on the known risks of allowing a depressive episode to go unchecked: disruptions in mother-child attachment and the increased possibility of more depressive episodes in the future.

Doctors who tell their patients to stop their medications "are giving out the best advice they believe exists, which is that we don't have a lot of data on the safety of these medications, and they don't want to assume the risk of having supported a patient in doing something that may have a bad outcome," Altschuler said. "So they're giving the best advice they know, but [often] it's the worst advice you can give for the mother in terms of mental health."

To Shari Lusskin, director of Reproductive Psychiatry at New York University, the problem is not simply that doctors don't know about every recent development in the field, but that they keep freely doling out authoritative-sounding advice anyway. That was a theme of many of the survey responses. "I went to my [primary care physician] for possible medications, but they knew nothing about which medications I could take when nursing, and basically gave me the choice to keep nursing and do nothing for my depression, or feel good using medications but stop nursing," wrote one mother. Lusskin herself cites a recent case—a woman who consulted her in her sixth month of pregnancy. She had suffered three episodes of severe depression and had been on medication when she became pregnant. Her first psychiatrist had told her to quit taking her pills, but had given her a prescription with six months' worth of refills. A second psychiatrist told her it was okay to keep taking the pills but to stop taking them just before delivery. Meanwhile, she was using her prescription from the original doctor to keep on her medications, and was completely baffled about what to do.

"She was getting conflicting advice left, right, and center," Lusskin said. "There are so many people out there who don't know what they're doing that it's scary. And it's not just the obstetricians who don't know what's going on. It's the psychiatrists, too. . . . Until doctors are better educated, patients will continue to get short shrift."

You might think that up-to-date information would be easier to come by in major cities, with ready access to teaching hospitals and research facilities, but this isn't necessarily so.

"I was in the D.C. area, with all these very good medical options, I went to a highly respected pediatric practice, and they didn't really have a lot of information on [antidepressants] and nursing," says Jeneva Patterson, a St. Louis mother who suffered the symptoms of PPD when her daughter, Sophie Manon, was born in 2002. "I was amazed at how little they knew. That was really surprising to me." When Sophie was only eight months old, Jeneva and her husband moved from Washington, D.C., to St. Louis, where they ran into the same problem. Eventually, Patterson concluded that Paxil was safe to take while she was breast-feeding—but only with the help of her husband, a research doctor in a totally unrelated specialty, and her brother, a clinical psychologist.

Other women find themselves in the uncomfortable position of telling their doctors what to do. Virginia Major, a Connecticut mother of two, has suffered with depression since her senior year in college. During her first pregnancy in 2001, she and her psychiatrist worked out a medication plan. "I had to tell my ob-gyn what to do," Major said, down to and including the exact dosage of medication to give her within one hour of delivery. "They deferred to me."

Other doctors respond by erring on the side of safety, or what they believe is the side of safety. Jeneva Patterson, for instance, went off her medications during her first pregnancy. During her second pregnancy, she had stopped taking them again.

We talked one afternoon over lunch in the dining room of her home on a tree-lined street in a gentrified section of downtown St. Louis. Eighteen-month-old Sophie ran in and out of the room in a diaper, babbling in a mix of French and English to her nanny in the kitchen and periodically appearing at Jeneva's chair to beg for Mommy's attention. Jeneva is thirty-six and strikingly attractive—blond and athletic, possessed of that pregnancy glow women hear about but few of us actually experience. But she has suffered from several episodes of serious depression, beginning when she was about thirteen, and the symptoms she described having after Sophie's birth—months of anxiety, crying spells, and irritability—fit the classic description of PPD, even though she never received a formal medical diagnosis. Regardless of that, her previous episodes of depression put her at high risk of suffering from PPD this time, and the social worker Patterson was seeing for therapy was urging her to restart her medications immediately, as a precaution.

"She wants me to go on meds and treat this aggressively, which I absolutely will not do," she said. Why not? I asked. "I don't think I'm bad enough," she said, even though just moments earlier she had been describing her recent bout with insomnia and anxiety, symptoms that for her are usually a herald of depression. "I'm not so depressed I can't function. And it's not something I feel comfortable about, with the development of the fetus."

Is her mental health at stake? Is she making the right decision? What should she rely on—her therapist, her doctor, or her gut instinct? I don't know, and neither does she. She looks at me, and in her clear gray eyes I see intelligence, determination, and more than a little fear.

CHAPTER FOUR

The Way It Is: Depression During the Child-Rearing Years

Being a mother with depression is hell. You know you're depressed. You know you're on a downswing so big you'll never get out, but you can't stop the swing. It just takes you with it. . . . It comes out in my body language. I get in my own way. I touch things before I touch them. Things get knocked over, I trip over nothing. I walk into walls. I just cry. I just shake. I just want to scream. I just want to jump out of my skin. Instead, I just go ahead and make dinner anyway. Just don't talk to me.

—LOLA EIJCKELHOF, THIRTY-NINE, OF BOISE, IDAHO,
   MOTHER OF ALEX, FOUR

Beckie Calvo, a thirty-eight-year-old mother from Burling-ton, Connecticut, suffered from postpartum depression (PPD) after the birth of her first child in 1992—but one day, just around her son's first birthday, she simply woke up well. "My hair was falling out, my usually clear skin was totally broken out, but I felt amazing!" she said. "I felt like I woke up from a bad dream!"

Beckie is a medical rarity: a person whose symptoms matched the textbook. Most experts define the "postpartum" period as ending at one year (although some limit it to six months). And, in this way, PPD often gets defined out of existence. If a woman is still suffering from symptoms like an inability to find pleasure in ordinary things, constant fatigue, or permanently disrupted sleep patterns at the end of one year— well, she supposes, this can't be PPD anymore. This must be "normal life." (Which is what Beckie had assumed: "I was convinced this was all just part of being a new mother. I was very wrong.")

But not everyone who suffers from PPD is as lucky as Beckie Calvo. (Although Beckie wasn't all that lucky, either: she suffered from untreated PPD three years later with her second child, too, though that lasted a mere seven months.) The worst symptoms of PPD can abate, only to be replaced by a kind of low-grade dysthymia that can linger for months or even years. Moreover, while PPD is a distinct form of maternal depression, not all maternal depression is PPD. In fact, *most* of it isn't. After all, depression is an illness that frequently recurs, some kind of triggering life stress can happen anytime, and hands-on child-rearing is, for most women, an endeavor that takes the better part of two decades.

So what does maternal depression look like in that interval between a child's first steps and the day he and his luggage get dropped off at the dorm?

The clinical literature on depressed mothers divides them neatly into two categories: "angry/intrusive" and "withdrawn." Snapshots taken from real life are more nuanced. The picture might be one of quiet hopelessness: Andra Baeten, who we will meet later in this chapter, sitting silently night after night with her nine-year-old daughter in the family room of their Chicago home, reading one self-help book after another in search of a path out of her postdivorce despair. It could be a picture of a mother choosing the Nuclear Option: Lola Eijckelhof in Boise, Idaho (who, along with the rest of the mothers I mention here, will show up later in the book), carefully counting out thirty-six sleeping pills—one for each year of her life—and swallowing them because that is the only way she can think of at that moment to free her husband and toddler son from the angry Monster Mom she has become. It might be a portrait of fear: Barbara Hinson (not her real name), weeping on the floor of her bathroom with the realization that the symptoms she has been trying to ignore are more powerful than her willpower. Or it might just look like stupidity: Amy Brownstein (not her real last name), sitting on a bench in a shopping mall in suburban Baltimore, crying because her four-year-old and her two-year-old are crying, too, and she needs to get them home but she can't remember how to work the latch on her four-year-old's car seat.

When a mother is depressed, the stakes are simply higher. "When I was single, the worst possible scenario was that I would lose my ability to make money and wind up on disability," I wrote in my journal a few years ago. That was a grim picture, but nobody really suffered in it besides me. But since becoming a mother, I wrote, any sickness of mine automatically affected three other people—my two daughters and my husband, who had to take up the slack for me whenever I was sick. Merely on a day-to-day basis, "the energy I use to keep going, to master my anxiety and just put one foot in front of the other is energy that otherwise I might spend on my children—playing with them, talking to them, listening to them." Which was all true enough—but now, looking back at that passage, I see myself hurtling straight into hubris. "I can't afford to be sick," I concluded.

Untrue, of course; if CEOs of major corporations are allowed sick days, the earth can spin indefinitely without my direct supervision. And yet it's easy to see how a mother could fall into this delusion, especially when her illness affects virtually every aspect of what goes on under her family's roof: the relationship between children and parents (and between spouses—a subject for another book), how much housework gets done and who does it, what the family eats, how much the television is on, how much of a social life the family has.

Depression affects every mother (and her family) differently. But there are some important underlying themes in the way depressed mothers behave and the forces affecting them.

Always in Flux

Motherhood is fundamentally a task of constantly redrawing boundaries: you start with a bunch of cells that is literally part of another person's body and, in theory at least, end up with two unique and independent adults. In between those two points, mothers (more than fathers, I would argue) must constantly negotiate the boundary between themselves and their children. This is a demanding task under any circumstances; it explains the never-ending stream of magazine features on "how to find time for yourself." For mothers struggling with depression, this constant task of drawing boundaries puts them squarely at the center of a fierce tug-of-war: the children's needs pulling them one way, their illness pulling them toward isolation.

The tug-of-war is at its height in the early years, when children's need for constant physical supervision is greatest. As one mother described it, "I ignored whatever I didn't really have to deal with. I let things get out of control, and then went ballistic over the level it escalated to." What usually happens next is that the mother feels guilty about her over-the-top reaction and withdraws. But as any mother who has ever attempted to, say, go to the bathroom with a toddler in the house knows, Mommy's attractiveness increases in direct proportion to her unavailability. Soon the child is back, demanding more, acting out more, doing whatever it takes to get her attention. And so it goes: irritability leads to withdrawal, which precipitates more irritability because the need to withdraw is being constantly thwarted.

"I went through periods that I just yelled all the time," one mother wrote. "Then I would be sad, cry, and not get out of bed. . . . [It was] a cycle of depression, rage, anger, guilt, sadness, and then all over again." Depression is every bit as implacable in its needs as any four-year-old with an empty sippy cup.

The Need to Withdraw

Phyllis Jean Spradley—PJ to her family and friends—sits in her immaculate kitchen, where the dark polished wood of the cabinets contrasts nicely against the blue-and-white striped wallpaper. The counters are virtually free of clutter, there are fresh flowers on the table, and her desk, which sits under a window overlooking a vista of rolling Oklahoma farmland, is bare except for a box of Kleenex and a family picture. I am a visitor, but I don't think that PJ went to a lot of trouble making her house immaculate in honor of my visit; I suspect that it's this way all the time. For one thing, two of her three boys are grown and out of the house, and the third is either off at school or down in the basement with his computer. For another, I recognize in her a fellow obsessive-compulsive— someone who needs order in her environment to compensate, just a little, for the periodic chaos in her head.

She is a slender woman of forty-nine, dressed in slacks and a neatly pressed red-checked shirt, her hairstyle sensible, her jewelry unobtrusive. Her manner is friendly but subdued; I don't get the impression that she laughs out loud very often. She has suffered from periods of depression off and on since puberty, and the illness runs in her family. Her mother suffered from depression (and does still, at eighty-two). Of the five girls in her family, she says, "three are on antidepressants now, and of the other two, she needs 'em and I do too." She laughs dryly. PJ would certainly be on antidepressants if they helped her, but the list of the ones that have failed to help read like a pharmacology manual: Zoloft, Paxil, Effexor, Prozac, Lexapro, Celexa, Wellbutrin. "Well, Prozac worked for me briefly, but then it bottomed out."

Of all the periods of depression PJ has known, the worst by far came when her boys were ages one, two, and six. She had just had a hysterectomy, and her doctor had put her on hormone therapy. This was in the 1980s, and PJ and her banker husband had just moved to a small town in northwest

Arkansas. The nearest state-of-the-art medical care was in Little Rock, several hours away. Her local doctor tried various combinations of hormones, different dosages, but nothing made things any better. The deep depression she fell into alternated between attacks of agitated rage and periods of "going into the deep, dark places," when all she could muster the energy to do was sleep or, at best, lie on the living-room sofa.

She refused to host playdates because she could not bear the noise and the disorder. Her six-year-old was old enough to play at friends' houses, she said, which provided him an occasional escape, but otherwise her self-imposed social isolation was almost total. "Friends or neighbors would come over and I just wouldn't go to the door. I wouldn't want to see anybody or have them see me." Her one-year-old seemed oblivious to her moods, she said, but her middle son, then two, cried a lot.

"It was just overwhelming, day to day, taking care of three little boys," she said. "Nothing pleased me. Nothing was ever good enough. Nothing ever went my way." She and her husband had dealt with infertility for years before they had children. Now, that was the worst part: knowing that the motherhood she had worked so hard to achieve was something she could rarely enjoy. "Kids are funny," she said. "They can be fun. But it happened quite often that I couldn't enjoy them just being children. If I didn't scream or yell at them, I just wasn't there. . . . And the more I screamed and yelled at my kids, the more depressed I would be, because I didn't want to be that kind of mother."

Who does? And yet summoning the energy to resist depression's undertow can be almost impossible. Many depressed mothers find themselves simply "checking out" mentally—performing the basic tasks of child care, meal preparation, errands, and laundry but doing these things automatically and numbly, with little emotional content. The voice in the backseat pipes up: "Mom, guess what I heard on TV?" and in a moment or two you hear yourself saying, "Uh-HUH," as if you have just heard something interesting, which you haven't. You may have just missed the last episode of *SpongeBob SquarePants,* or you may have missed a clue that your child is being bullied at school. Problems the mother does not hear about she can have no hand in helping her child think through, and there are other things left unsaid as well. "I was less affectionate, less praising, less communicative, not involved with current events," said one mother about her episodes of depression. Another described herself as "mechanical about [my children's] care. I did everything I was supposed to do, but had tremendous difficulty playing or engaging them directly. Sometimes I avoided eye contact." A mother with older children (ages ten, thirteen, and fourteen) said that because of her depression, she would "rather [lie] in bed and watch TV" than interact with her children. (She added—wistfully, I thought—"I hope they know what is important and will ask me if they need help.")

I hoped so, too, but I was doubtful; children always notice emotional disconnection, though with teenagers it may not always be immediately apparent. Younger children, however, often pay an immediate price for their mother's illness: outings, exercise, playdates—all tend to get canceled or never scheduled because Mom can't work up the energy to go anywhere, or the energy to maintain appearances. As one mother wrote, "[I remember] not wanting to get myself dressed and take [my son] to the park, or to special activities that I had signed him up for, being afraid to make playdates [or] then canceling playdates." It is hard to make firm plans for next Thursday if you are not sure if next Thursday will be a "good" day or one of those mornings when you can barely crawl out of bed.

What interferes with getting up? It's not necessarily existential angst, or pondering the meaning of life. It's flat-out physical *exhaustion*. Depression wears you out. There may be no outwardly apparent stressor, but the brain keeps the body revved up the same way a car with a defective idle mechanism can be sitting in the driveway and still have an engine roaring at 2,000 revolutions per minute. Sometimes this manifests in insomnia; sometimes it manifests in disturbed sleep patterns that steal your rest even while you are, technically speaking, asleep. There's a big difference between a night spent in restless, vivid dreams and a night spent in deep, brain-reparative sleep. (After too many years of navigating my way in and out of depressive episodes, my sleep patterns are permanently askew. Under certain conditions, I can sometimes start dreaming before my eyes are shut. On our honeymoon, when I was exhausted by the previous week's wedding events, I fell asleep while sitting up, eyes wide open, in the dining room of our hotel in Tortola—a fact my husband realized only when I said, out of nowhere, "Why doesn't he get his own taxicab?")

Any mother dealing with this will have to get through the day while constantly resisting the siren call of the bedroom. The desire to escape can be intense. "One time, in an awful episode, I just locked myself in my bedroom," one mother said. "[My children] slipped notes under the door. [The oldest] was able to write but [the youngest child's notes were] just in scribbles." Some lash themselves to the mast by inventing a million errands to run: stash the kids in the backseat, let's go meet the Easter Bunny!—and then, Macy's is having a sale! It's easy to justify; kids always need to be entertained (don't they?), and there's always something to buy. For that matter, there is always work waiting at the office. One mother I heard from described a three-year-long depressive period in which "I worked as many hours as possible to avoid going home. . . . I felt that there was already so much being demanded from me that I had nothing left for them. On many days, I felt that they were better off without me."

Mothers who don't feel up to leaving the house can always invent more housework—and by "inventing housework" I don't mean doing the laundry or unloading the dishwasher. I mean starching and ironing the bedsheets, or probing for dust between every single Venetian blind, or scrubbing even the nonmildewy spots of the bathroom tile with a toothbrush. "I was a perfectionist," said one mother in the survey. "I had my children put their toys away in alphabetical order."

It's just a theory, but I've never met a mother who was a neat freak whose depression was not characterized mainly by anxiety. When there is disorder in the mind, the need for order in the environment can become intense—which is unfortunate for children, since children are (and should be, at least during playtime, the antithesis of order). During my bad times, I wake up every morning with a low hum of anxiety in my head, and, when things are at their worst, by 4 P.M.

I'm so spun up I feel as if I could glow in the dark. This happens so reliably that I've given this phenomenon a name: the Four O'Clock Willies. At such times, tackling projects like organizing the video collection or sorting through my daughters' Barbie stuff are the only things that can keep me from bouncing off the ceiling. But the kids have to be somewhere else.

Other mothers manage to fight off the urge to call a time-out and go to bed, only to collapse on the living-room sofa. "I find myself on the couch totally withdrawn while [my children] play together," one mother wrote. "They are two and three. I won't even hear what they say to me unless they approach me and get in my face."

Even when the need to withdraw does not result from physical exhaustion, depression can make a mother shrink from physical contact—not so much from the desire to avoid physical touch (which can be soothing), but out of the fear that touch expresses a need she is unable to answer. She is like the kid slouching in the back of the classroom, trying desperately to stay off the teacher's radar screen. Anything's better than having to say, "I don't know" or "No, I can't." One mother wrote, "I have had days during which, at the end of the day, I realize I have not hugged my oldest child at all." Another said, "I find myself being exasperated or cringing when they all clamor or cling, wanting hugs and kisses and affection when I just feel numb and useless." That numbness sometimes makes it hard to realize just how disconnected you are. One mother hospitalized for depression said that during her stay, doctors made a videotape of her with her children. She watched it months later, after she was feeling better, and described what she saw on the tape: "I just sat there with them. Nothing else. No interaction or anything. I was shocked."

Some mothers I heard from adopted the simple expedient of sending their children to their rooms so they could be alone. "I will pick him up from school and we will come home and I will send him to his room to go play," one mother told me. "Then I will sit down and watch the afternoon talk shows and whenever he [comes] downstairs I will send him back to his room and [tell] him that the program on TV is not for him." If all else fails, there is always television. "Sometimes I just prayed that my child would keep watching TV so I didn't have to interact with her," said one mother. On the other hand, as another mother pointed out, television may not always be the worst option: "Mr. Rogers is a lot nicer than I am when I am depressed."

There are ways to cope with exhaustion and the need to withdraw, which I will get to, but the important point to remember here is that the need and the exhaustion are *real*. Caffeine may help temporarily but it won't cure your fatigue, and it does no good to berate yourself for being tired "for no reason." You have a reason: you are dealing with depression.

Our Lady of Perpetual Irritability

One of the questions on the survey asked mothers to describe "all the little and big ways your depression affected how you behaved toward your children."

"I just asked my eleven-year-old son this question," one mother wrote, "and his response was, 'You chewed our butts.'"

The popular perception of depression is that it is makes

people sad. Chronic irritability is a less recognized but equally common symptom, and it can escalate up to anger attacks—periods of uncontrollable, hysterical rage. Sigmund Freud's famous dictum that depression "is anger turned inward" may be true in some instances, but when it comes to maternal depression, that rarely seems the case; for depressed mothers, depression is anger turned outward—at the kids.

"I talked in a strained voice or slammed my bedroom door [or] overreacted to daily mishaps," one mother told me, and another said: "I might look at one of the kids in their room and, instead of saying a nice greeting, I will bark, 'Why haven't you cleaned your room?' " At other times, anger takes the form of putting inappropriate expectations or responsibilities on a child who is too young for the job.

"When he was learning to read, probably around five or six years old, I would be very short-tempered," one mother wrote. "I expected a five-year-old child to have the attention span of an adult, and when that didn't happen, I would explode. To this day, my son doesn't want to read anything and I believe I am to blame for it." One mother of a four-year-old girl wrote, "I start to expect her to have the manners of an adult. . . . With my twenty-month-old, I oftentimes don't watch her as carefully as I should and then get mad when she gets into the markers and makes a mess. I sometimes blame her big sister for not watching her and put the blame on her, when in reality I was upstairs with a pillow over my head."

Sometimes it's the little things—literally. Like a little person in the backseat asking the same question over and over, or tapping the car window out of boredom. In fact, child boredom and adult depression are a volatile mix. "Constant singing or nonsense noises in a closed space (riding in the car) makes me very irritable," one mother told me. Loud play, boisterous behavior at bedtime, kids who yell for jelly beans when they should be eating breakfast, a child who has to be told six times to brush her teeth—and, suddenly Mom is screaming. "When their room would be messy, I would go in and end up yelling and screaming, dumping dresser drawer contents on the floor, throwing things around, making the mess much worse. It was horrible," one mother told me. Another said, "Sometimes they just look at me like, 'Who is this in my mommy's body?' I come to a point when it gets out of control and I just have to walk away."

Depression leads to irritability, which leads to anger, which leads to screaming, which, sometimes, leads to abusive behavior. "I know that I get it from my father but I often find myself yelling when it just isn't necessary," wrote one mother. Another mother confessed to yelling, slapping, and shaking her son, but added, "I never told him he was worthless and no good like my mom told me." She sounded defensive, which was understandable. I was frequently amazed at what mothers granted anonymity would freely confess:

"When my middle daughter was about three years old, I got so angry at her for falling out of the car because the door was not closed properly that I kicked her when I got her home."

"I'm a name caller. I'm also very nasty and try to make [my children] feel dumb."

"When the angry spells hit, then I will yell, 'Shut up, you talk too much, and all you are saying is stupid.' "

"When he was eight, I told him he was a jerk."

"My [nine-year-old] daughter was upset that her dad (we are divorced) couldn't spend more time with her, [and] then she tripped and instead of holding her and consoling her crying, I yelled at her for being so clumsy."

"On my son's thirteenth birthday he was very wound up and wasn't behaving well. Instead of taking him aside, I continually disciplined him through the party in front of his friends," one mother wrote, adding, "I'm so ashamed that I didn't handle that more maturely and [that I] hurt my son."

Irritability, just like withdrawal, frequently leads to social isolation for children. "I have been crabby and disorganized to the point of near chaos in the house," one mother in the survey wrote. "It affects [my children] because they can't always have friends over and the stress level is high with all of the clutter and disorganization." At other times, the social isolation was the result of marital tension resulting from the mother's depression. "The house remained messy and this created friction between Mommy and Daddy," one mother wrote. "Few friends were involved in [my son's] life."

Sometimes the social isolation is the child's choice: better not have friends over at all if you never know what mood Mom is going to be in that day. "I think the biggest thing that was hard for my kids was how unpredictable I was," one mother told me. "If I was feeling pretty good, I was easier to get along with, less moody, more able to respond to them spontaneously, more able to entertain them and their friends. If things were not going well, a simple request could be met with a short angry response from me. I was less able to be with them, to have fun; [I was] more rigid about friends, schedules, etc. They were never sure what to expect."

"More rigid" is a telling phrase: one of the hallmarks of depressive thinking is its either-or nature. Solutions do not occur, other data do not filter in, because the mind is in lockdown mode. Your son leaves his dirty clothes on the floor and this fact immediately becomes evidence of a global plot to enslave you with housework; you don't notice that he is absorbed with the fact that he had a huge fight with his best friend that day. Later, though, things look different. "I sought help with antidepressants when my children were seven and nine," one mother told me. "Prior to that, I felt like I was constantly in a state of irritation with everyone. I felt like my nerves were raw and every normal childhood behavior rubbed [me] the wrong way. I realized how sick I'd been after the meds kicked [in] and suddenly my children's behavior looked completely different to me."

The Need to Conceal

For mothers, admitting to suffering from depression is admitting that having children has not brought them the kind of serene fulfillment that Hollywood moms talk about. Then again, the word *depression* has been used for so long in so many nonclinical contexts that its medical meaning has become lost in the static. It's not surprising, then, that between the social stigma and our inadequate vocabulary, many mothers find it difficult to come out of the closet. The choice, usually, is between way too much information and way too little. "What's the matter?" someone asks, and what you might say if you were being excruciatingly honest is, "My joints ache; I want to sleep fourteen hours a day; I have no appetite or libido; I cannot read or do a crossword puzzle; I am trembling with anxiety; I cannot enjoy any ordinary pleasure, and I suffer from the demonstrably untrue belief that my life has been one long disaster." Instead, you say, "I am depressed." If you are lucky, you will be talking to a person who understands what you mean. If not . . . "Cheer up!" they say, and walk away thinking, *What a wimp. What does she have to be depressed about?*

That, at least, is the fear, and it helps to explain why so many people take antidepressants in secret. Not being open about our illness then contributes to lack of understanding— and so the cycle feeds itself. Of the 332 mothers in our survey who answered the question, "Do you think that your friends and family, for the most part, consider your depression as serious a medical problem as other chronic medical conditions—high blood pressure or diabetes, for instance?" only 23 percent said, "Yes, definitely."

But in defense of family and friends, it's hard to realize the seriousness of a condition that is kept under such heavy wraps—and most of the women I heard from said they tried to conceal their illness from their husbands at least some of the time. They didn't want to burden their husbands, some said, while others said their husbands simply could not understand that depression was a bona fide illness; some just admitted that they were ashamed. "I did not want my husband to see me as weak or mentally ill," said one. Another wrote, "Sometimes I worry he'll think I'm crazy." Given this reticence, it's not hard to understand why fewer than half of the women in the survey said that their husbands or partners routinely took over some or all of their child care or household duties when they were sick. "He kicked in every blue moon," one mother wrote. "Other times, he thought I was just a crazy black female with an attitude." Another mother said, "He will feed them, he gets them ready for bed, he takes them for walks, to the park. He just won't do anything to help *me*." It could be that her husband is insensitive to her needs—or it could be that he's not a mind reader.

As in the family, so with the outside world. One mother told me that she hides her depression by "going to church and putting on the big cheesy smile, and distributing hugs to everyone in the congregation." (Every church seems to have at least one of these.) Another wrote, "I would get all cheery and just laugh for no reason and that would make my daughter laugh." But putting up a false front takes a lot of energy. "I'll fake smiling, pretend I'm listening to [my child], just say, 'Yes, okay,'" one mother wrote. "[I don't have] the energy to answer all the 'whys.'"

Diversion is another tactic. "I try to play more with them or take them out somewhere to get treats—anything to divert their attention [from] me," the same mother added. "My son has seen me cry and got very upset. Since then, I cry in the bathroom into a towel if I have to." Bathrooms are the preferred spot for weeping—you can turn on the water there—though closets and bedrooms got mentioned frequently, too. And many of the mothers, in addition to these tactics and others, took the tried-and-true route to short-term happiness: alcohol, prescription drugs, or illicit street drugs. "If I used alcohol to self-medicate, I would be capable of acting interested and happy and energetic for a short period of time."

Many women justified their secrecy by saying they wanted

to protect their children. But children always know at least twice as much as we give them credit for, and they're also good at spotting phony behavior. It's appropriate not to expect children to be little adults, or weigh them down with adult cares; it's not appropriate to enroll them in an ongoing deception, or lead them to expect that life will consist of treats every day. "I'm sorry, sweetie, but Mommy's not feeling well" is not going to warp a child for life, even if you have to say that sentence more often than you'd like. A straightforward admission of sickness might even relieve a child of the childish fear that he or she is the cause of your unhappiness, and give the child the chance to display a little kindness. (More about that in chapter 8.)

What Are the Rules Today, Mom?

Any parent who takes the job description seriously can attest to the fact that being consistent in making and enforcing rules is one of the hardest aspects of what we do. Making rules takes self-confidence; enforcement takes energy. Mothers with depression find both qualities in short supply. "At times I would be too tired to enforce anything," one mother told me. "I just wanted to go to bed and sleep, so as long as she wasn't in danger, I let her do whatever she wanted."

"I don't enforce rules," one mother said flatly. "I feel that everything is pointless, that I am not [of] value to listen to and be heard." Another wrote, "At times I don't trust in myself enough, and second-guess my decisions about discipline." The rationalizations start; after a while, a certain fatalism sets in. *What's the point? You won't win this one. After all, you have to pick your battles. I'll let it slide. Next time . . .* "When my son was eleven he would not do his homework when asked or would do it poorly and refuse to redo it," one mother told me. "Rather than get in a battle I couldn't win, I gave up and quit asking to see his homework or do anything about it. I didn't feel I had the strength to deal with the problem, and I felt my inability to control his actions reflected my inadequacies as a parent."

Other mothers slacked off on enforcement because of guilt. "To compensate because of my illness, I know that I deliberately excused behavior that normally would not be tolerated from my children," one told me. "I had everything I could do to just get through the day without a collapse of my own."

The problem of setting limits often comes up specifically in the context of nutrition, which, considering the billions of advertising dollars spent encouraging kids to eat salt- and sugar-laden snacks, is not surprising. One mother wrote that her episodes of depression cause her to "not [make] enough effort to limit snacks and things that aren't healthy foods, and allowing the junk food just to keep him quiet." Another mother, who had a teenaged son, wrote, "I stopped cooking dinners when my youngest was sixteen or seventeen. I ate very little and he had to learn to cook. I know he wished his mom would cook like his friends' moms and he wanted to invite his buddies over to eat."

Kathy Johnson is a thirty-nine-year-old stay-at-home mother in Lithia, Florida. She has dealt with depression off and on for most of her life, and feels that now she has her illness under control. But she thinks it has had a direct effect on her nine-year-old son's eating habits. "My child is a very picky eater," she said, and when he was a toddler, depression sapped her energy for "continually trying different foods, and making him try them over and over." Today, she says, "I'm faced with a child who has a very unhealthy diet. . . . I think if I'd had the gumption to deal with it when he was littler, I might not have to deal with it now."

The effects of this inconsistency on children varied, but, as one might expect, none of them were good. Several mothers made explicit connections between their children's confusion and rebellion and their own failures to apply rules consistently. "I think it confuses them," one mother wrote. "It makes my oldest disrespectful, because he feels like he can't do anything right." Another described how the general lack of limits set the stage for increased sibling rivalry at her house: "Chores weren't done regularly and their attitudes toward each other [were] more critical. They seemed to fight more often and had less tolerance of each other's needs."

Regrets

The women I talked to were not poor; with a handful of exceptions, most were married to a wage-earner and/or comfortably middle class. But even in this relatively affluent group, the subject of financial problems created or exacerbated by depression came up frequently. Depression is expensive: mental health professionals rarely charge under $100 an hour, and a month's supply of some of the newest antidepressants (Cymbalta, for example) retails at more than $400 for a thirty-day supply. Without health insurance, a lot of medical care is simply unaffordable; with it, it can still be prohibitively expensive. Then there are opportunity costs—the concessions on employability imposed by chronic depression: missed workdays, used-up sick leave, the promotions missed, the better-paying jobs turned down because the added stress would be a killer.

"I find I get overwhelmed easily. I cannot take on very many ventures at once. . . . I tire easily, too," wrote one mother, explaining why she doesn't hold a paying job. "This means financially I cannot buy the children all the things I would like to. I've had to say, 'No, we can't afford that.' I then feel guilty afterward."

"I have to say that the biggest effect [on my daughter] was my inability to hold a job due to my depression," wrote one mother, a forty-eight-year-old attorney. "I have trouble dealing with money when I'm down, and she has had to live through several episodes of having our utilities cut off during those times. . . . I have no real assets and no romantic partner, so I feel that I haven't given her the life she deserves."

Money, then, is one regret; time is another. Depression steals time, and children offer us a time-lapse picture of its passage: the sweet round face of toddlerhood dissolving into the grown-up six-year-old; the gawky fourteen-year-old who one day, in a surge of testosterone, sprouts a beard. "I miss the hopeful, excited person I was when I was younger," one mother told me. "My children don't really know that person and that makes me feel very bad, like they are missing out on a part of me that I really love."

I understand her feeling; my children have missed out on parts of me, too. "My great fear is that this sickness will rob me of enjoying what should be an idyllic time," I once wrote in my journal. "Every time I think I'm rid of it, it shows up again. It's like I've turned around and there it is on my back, a fat black tick, looking at me with lazy, insolent eyes: *You thought you could get rid of me?*"

So when my four-year-old says, "Mommy, can we go catch lightning bugs?" I say yes. "Yes" is my default option; there has to be a really good reason to say no. Even if I don't feel like chasing lightning bugs at that moment, it is a kind of mental discipline to stop what I am doing and focus on the moment. I am acutely aware of the time I've already missed.

And—not always, but much of the time—the moment that my kids win when they pry my fingers off my to-do list turns out to be a gift, a package waiting to be opened. Other mothers described the same thing. One told me about a day when she was exhausted and not feeling well and her son was misbehaving. "[He] wasn't listening to me, wasn't doing what I asked [and] I was ordering him around. [He] was whining constantly, so I put him in a time-out. Then I realized: he wanted his mom. So I dropped everything I was doing and focused on him. It was amazing how his demeanor changed in a matter of minutes."

"My kids were a life raft," said Saranna Thornton, a forty-five-year-old economics professor at Hampden-Sydney College in Virginia. She has four children, and a history of depression that began when she was eighteen. After the birth of her third, in the midst of a "crashing" episode of PPD, she found solace in her children—breast-feeding the new baby, snuggling up at night with her two older sons, then seven and four. (Her husband was in a graduate program in another city, and by then had gone back to school.) The physical contact

was healing, she said—"sort of analogous to puppy therapy."
Medication helped—was essential, in fact—but while she
was waiting for her brain to heal, human touch was a wonder-
fully effective salve.

It is a scene I like to imagine: mother, baby, and two little
boys, one big heap of arms and legs, the smell of talcum pow-
der and small dirty feet, the unexplainable calming effect of
beloved skin against your own. Such moments make up for a
lot of regret.

The challenges facing mothers who deal with depression are
so daunting it's tempting to conclude that motherhood and
depression are just incompatible. Historically speaking, many
people have thought so: up through the 1940s, there are nu-
merous first-person accounts of women whose depression
was treated simply by being put in an institution for decades.
And of the women I heard from who were single mothers,
several mentioned their fear of having a disgruntled former
spouse use their illness against them in child custody pro-
ceedings.

But motherhood and depression are not incompatible; depression is just one of a thousand complications that may at-
tend the job of parenthood. Any woman with a history of
depression should take that fact seriously when planning to
have children, but women with all kinds of serious health
problems become mothers—women with diabetes, hypertension, multiple sclerosis, hearing impairments, carriers of genetic defects, paraplegics . . . the list goes on. Not only does
society rarely question their wisdom in deciding to bear and

raise children, for the most part it applauds them for their courage. For that matter, even women who become mothers in robust health have no guarantee that their health will last forever; life is a cosmic roll of the dice. In some ways, the struggle with this illness can enhance parenting skills or create skills you never knew you had (a subject we will get to later). The only requirements are recognizing the problem and learning how to manage it.

Easy to say, not always easy to do. The first and most basic task is internal, and that is recognizing that you are sick—that what you are experiencing is *not* normal, that your inability to cope with daily life is not the problem but the symptom of a real, and treatable, medical condition.

The problem, of course, is that mothers are notorious for being so busy taking care of others in the family that they forget to take care of themselves. I was struck, for example, in reading a 1976 biography of Sylvia Plath (who, along with Anne Sexton, is one of the twentieth century's most famous depressed mothers), by the number of utterly mundane crises that beset her in the last days before her suicide. In 1963, the year she died, London suffered its worst winter in 150 years. So in addition to the fact that her husband, poet Ted Hughes, had just left her for another woman, Plath was dealing with problems finding child care for her two small children so that she could work, power failures, frozen water pipes that burst and discharged wastewater into the bathtub and temporarily cut off her flat's drinking water, influenza that hit her and both children, and—the final insult—chronic sinusitis. The weather-associated problems alone lasted for weeks. Throughout, Plath devoted far more time to taking care of her children than she did to getting medical help for herself. The last thing she did before her suicide was to set out milk and snacks for her children, for when they woke up.

And yet the moment of recognition does come. For Liz Bunker, the "nonworking" stay-at-home mother we met in chapter two, the realization came when she went into a tirade at the receptionist at her children's school one day over a trivial issue. She had suffered several episodes of depression by then, but had recently stopped taking her antidepressants in the hope that, if she just ignored them, her symptoms would disappear. After her outburst, she went home and pulled the bedcovers over her head. "I just wanted to hide under a rock," she said. But it was a wake-up call: she went back on her medication. Between that and exercise, she said, she's been fine ever since.

For Andra Baeten, depression was a kind of slow-motion train wreck, and the recognition that she needed to do something about it came almost as slowly. In retrospect, she thinks she was mildly depressed during much of her fourteen-year marriage, but things became a crisis when in 2003 she told her husband she wanted a divorce. (Her ex-husband is "a great guy," she said, "but it was a marriage of friendship, not of love.") The parting between the two of them was sad but not bitter. The person who suffered the most was Andra's eleven-year-old daughter, Ashley, who blamed her mother for breaking up the family. The finger-pointing was unnecessary: Andra was already heaping blame on herself. Ending the marriage was a relief as far as she was concerned, but the effect it has had on Ashley is something "I am going to regret for the rest of my life." A few months after the divorce, Andra's mother died unexpectedly of a massive heart attack, still un-reconciled to Andra's divorce. For her, 2003 will always be the year when "my daughter lost her father, I lost my mother, and my father lost his mind."

It was a bad time—"bad, bad, times five," she says. Some days, she felt completely numb, except for the pounding of her heart. Her brain seized up with anxiety. She was doing well to remember how to run the microwave, much less how to get Ashley off to school every day in clean clothes with her home-work done and a lunch packed. Andra has five siblings, most of them nearby, but Andra was the one her grief-stricken father called the most often in the weeks after her mother's death, at all hours of the day and night. The more he leaned on her, the more she felt the ground under her feet crumbling.

She went to her priest, seeking comfort and advice. "The priest said, 'You know, you really need to get help for your dad,'" she said. "'It'll be okay'—that's pretty much what he said to me. . . . My faith just up and left." At night, Andra said, she would lie in the dark beside her daughter, trying to give her words of comfort and guidance and feeling, in her words, "like a complete fraud. . . . How could I tell her what to do [on] any given day when I couldn't get my own life together?"

A friend who worked for a local neurosurgeon told Andra she really needed to see a doctor, so she did. He gave her a prescription for Wellbutrin and a referral to a therapist. After a week on the pills, she looked at the package insert and read all the fine print on possible side effects. "Dear God, *no,*" she thought, and threw the rest of the pills away. But at least psy-chotherapy was helping, a little: "The best thing she told me was that I wasn't going crazy." But the visits were expensive.

With the divorce pending, she was still covered on her husband's policy, but she was unimpressed with the in-network providers she was offered, and the policy covered less than half the cost of the therapist she was seeing. The out-of-pocket cost for therapy came to $125 a week, which was just not in the budget.

So she began haunting the library, bringing home piles of self-help books. In the evenings, she and Ashley sat in the family room together, a thousand miles apart. Andra would read; Ashley would watch television. After an evening spent in silence, Ashley would go to bed. Andra would stay up reading until the small hours, trying to figure out what to say in this conversation she needed to start with her daughter about the turns life was taking, looking for guidance, hope, advice—any little piece of wreckage to hang on to, really—and finding nothing. For months, that was their life. They had no visitors. "I didn't have the patience for her to be here with friends tearing the house apart, or asking me to do things like make popcorn," Andra said. "We [didn't] do things—go for a walk, bike ride, go to a museum, like we used to do. . . . It was a solid year of sickness."

She told me this story one day as she sat at her kitchen table, dressed in blue sweatpants and a gray Abercrombie and Fitch sweatshirt, smoking nervously. It was early afternoon, and the house was quiet. The kitchen where we sat had an unlived-in look, as if nobody ever cooked there. Andra offered me some coffee cake from a local bakery and pushed a slice of her own around on her plate as she talked. She was a small-boned woman whose movements were quick, almost birdlike, and when she smiled, she was—briefly—beautiful.

But she was pale and thin; it was clear she had not been eating coffee cake, or much of anything, lately.

On the day we spoke, her recovery was still in a tentative stage. Things had began to turn around, Andra said, when she finally became desperate enough to tell some friends about the extent of her despair—or, as she put it, "allowed myself to let my friends be good friends to me." She was shocked at what she discovered: there were all kinds of things going on under the veneer of ordinary life. Several friends had survived similarly bleak periods and had some wisdom to impart. "All you have to do is breathe," one told her. Another friend was about to undergo surgery for a brain tumor, a crisis that helped Andra put her own problems into some kind of perspective. "I began to recognize that life rewards action," she said, so she took a couple of classes at a local college and joined some local women's groups. Slowly, she began to feel better.

"I'm not in a place where I can feel confident yet," she told me, but she is working on it: exploring new spiritual paths, trying to repair the lines of communication with her daughter. Ashley is her main motivation to get better. "Thank God I have her," she said. "If I didn't, I'd probably still be sitting here with my stacks of self-help books." But she fears their relationship has been damaged. "I don't think she has the confidence in me anymore when we try to have talks, and that hurts, because she used to." Instead, Andra writes letters to her daughter every week. Even if her daughter doesn't read them now, or doesn't understand them, Andra hopes that someday she will. After all, the clock is running out on Ashley's childhood. "I'd better get my act together somehow, some way, because she's going to be a teenager."

CHAPTER FIVE

Rats, Monkeys, and Mothers— or, Will My Children Inherit My Depressive Genes?

Continuing to live—that is, repeat
A habit formed to get necessaries—
Is nearly always losing, or going without.
It varies.

This loss of interest, hair, and enterprise—
Ah, if the game were poker, yes,
You might discard them, draw a full house!
But it's chess.
—PHILIP LARKIN

*L*ola Eijckelhof was holding her son, Alex, in her arms one day when he was less than a year old when it became vividly apparent to her that even a small baby understands when something is wrong with his mother's emotional state. Lola had been profoundly depressed since Alex's birth, and as she looked down at her baby that day, she said, "I will never forget recognizing that what I saw in his eyes was fear—fear of me. It shocked me. It made me ashamed. This was my child and there was no sparkle in his eyes when he saw me, there was no eagerness when I picked him up. . . . What a wake-up call."

The postpartum depression that descended after Alex's birth in 2001 eventually morphed into a state of permanent major depression that lasted for the better part of Alex's first three years. Lola tried medication and therapy, and they helped—but only so much. Her depression manifested itself not in withdrawal or fatigue but in a constant state of irritability. She lived on the razor's edge of rage. If Alex made waking-up noises in his crib, "I would fly up the stairs to get him, cursing all the way at the top of my lungs," she wrote—and then, somehow, pick him up gently (and, sometimes, tearfully). She never abused Alex, but the simmering anger came through in everything—her body language, her silence, the way she slammed the kitchen cabinet doors. She would make six different decisions about whether Alex was warm enough at night, and then the minute she crept into his room to check his covers, he would start to stir; she thinks he felt her anxiety even in his sleep.

One night in July 2002, she hit a wall. "No matter how hard I tried to be good, to be calm, to stay on a consistent emotional level, it just wasn't happening, even with medication and therapy," she said. If it was torture to live, she reasoned, it had to be torture to live *with* as well. If she weren't around, at least "her boys" would be free to start over with a new wife, a new mother. What she did next was not a conscious decision, she remembers; it was just automatic, in the same way her hands would start picking up dinner plates at the end of a meal. She went through the house and cleaned it thoroughly; whatever else people might think of her, they would at least know she kept a neat house. Then she wrote a note to her husband and swallowed thirty-six sleeping pills.

In the standard made-for-TV movie, this would be the plot hinge—a dramatic rescue, perhaps, or a life-altering epiphany. In real life, Lola woke up in the hospital, angry that she had failed even at failure. The next few months were only slightly better; she changed medication, resumed work in therapy— but all along, she was acutely conscious of the link between her worst moods and her menstrual cycle. Finally, in early 2005, after extensive research and thought, she made the radical decision to undergo a complete hysterectomy—uterus and ovaries. She was thirty-eight. She woke up from the surgery knowing that something was very, very different, in a good way. "It was like having a weight lifted off," she said. Since then, "the depression [has been] just rolling back. I'm going through menopause, which I love, except for the hot flashes. And it's like, here I am—I'm back." As for Alex, she said, "It's very strange. I have a kid for the first time. He's responding to me. It's like falling in love with your own kid." For example: "He wouldn't look at me, really *look* at me, until recently. I mean, look in my eyes. He would look off to the side of me or above me. Now he takes my face and looks at me."

Today, Alex is "a happy, loving child," which Lola credits almost entirely to the consistent and devoted care he has always received from her husband. Being exposed to her illness has made Alex an extremely sensitive little boy, she said, but that's both gift and curse: at four, his tendency is to assume that he is the reason for the moods he detects in others. In the predatory world of four-year-olds, this can be a huge vulnerability. The repair work, Lola realizes, will have to focus on teaching him that not everything other people do to him is a commentary on his personal worth. "He's going to have to build up a shield." In ways she cannot see yet, Lola fears, her illness has left an imprint. "I think he'll carry it for a while," she said. Maybe—who knows?—for life.

Lola's fears mirror the feelings of many of the depressed mothers I interviewed and heard from. For the more pessimistic ones, fear would be a relief; it would imply that maybe something could be fixed, or at least salvaged. In their perception, their performance as mothers has already been a disaster, the effect on their children a done deal. "My daughter deserves a mother who is not depressed," one mother told me. "It's in their genes," another said. "It has cheated my child out of the mother I could have been if not for the depression. This is time we can never recover," was another fatalistic comment. Another mother said she fears "that all three of my kids will have severe marital difficulties, and all will definitely need therapy in their lives."

The sense of resignation here reflects the informed layman's current understanding of depression: one, that it's a biochemical malfunction in the brain (and so by definition tough to treat), and, two, that the tendency toward depression is genetic. Both of these statements are true, as far as they go.

But the pace of research into the biochemistry of mental illness has moved so fast in recent years that today, those statements don't even begin to encompass reality. The truth is, even if your children inherit your depressive genes, they are not doomed to suffer.

The most recent research shows that the heritability of depression is part of a complex dance in which our genes affect our environment (a counterintuitive notion) and our environment affects our genes (an even more counterintuitive idea). What's more, this dance continues throughout life, its pattern subtly altering with every passing year. We may be stuck with the eye color we were born with, and we may suffer traumas we cannot control, but our brains never *stop* changing. The poet Philip Larkin was wrong: life isn't chess—it really is poker, after all.

In fact, the outlook for people who suffer from depression has never been better. Today, researchers don't just know that depression is some unfortunate collision of genetics and biology; they're beginning to figure out which genes are interacting with exactly which kind of environmental cues. This is research that could someday soon yield treatments vastly more effective than the ones we have now—specifically, drugs that can stop the cascade of neurobiological misfires in the brain at their source, rather than treating the imbalance of various neurotransmitters caused by those misfires (which is what current antidepressants do).

More important, scientists are also learning about the timing of brain development and when certain regions of the brain are most vulnerable to being "wired" for depression. One such window may be as early as the first trimester of pregnancy; others are open in the neonatal period; a few remain open up through adolescence. Understanding how this works will eventually lead to better treatment for depressed mothers during and immediately after pregnancy, and that will translate into better health for their babies. It also helps us figure out age-appropriate therapies aimed at helping high-risk children learn coping behaviors that can help to buffer them from stress: optimistic ways of perceiving the world, resilience in the face of adversity. The latter is not some pie-in-the-sky grant proposal; Martin Seligman and his colleagues at the University of Pennsylvania have been field-testing such programs in elementary and middle schools for several years now, with promising results.

These are things you would not know unless you subscribe to publications like *Archives of General Psychiatry* or *Neuroscience and Biobehavioral Reviews*. Research in this area proceeds in small, painstaking increments in places the public never sees, and only the most spectacular discoveries find their way into the mainstream print media (where it even more rarely makes the front page) or broadcast outlets such as National Public Radio. But the truth is that when it comes to depression, there is a great deal within our control—both in terms of our own health (which directly affects our children) and our children's.

As recently as 1970, the brain was to neuroscientists what America was to the early Vikings: a virgin territory that promised fabulous discoveries—if only somebody could figure out how to get there. Science has come a long way since then, and the pace of discovery is quickening.

The woman who made maternal depression a topic of serious scientific research strides into her office overlooking the Hudson River, on the fourth floor of an ugly redbrick building known as the New York State Psychiatric Institute. Myrna Weissman has just come in from walking four blocks in the rain, but in her fashionably severe black suit she looks remarkably unlike the stereotype of the frumpy woman scientist.

Weissman's field is epidemiology, the study of how disease spreads in a population, what environmental conditions affect its course, and what measures can be used to control it. It's a field most people think of as relevant to the study of contagious and/or catastrophic diseases like measles or the Ebola virus—not psychiatric ailments. Up to the late 1960s, psychiatry was still the realm of various schools of academic theory, all of which viewed the human unconscious as a Dalí-esque landscape of half-buried conflicts, forgotten trauma, and forbidden sexual urges. But a small number of neuroscientists were beginning to reexamine an idea that Sigmund Freud had considered, though he had lacked the tools to explore it: that depression was some kind of physiological disorder of the nervous system. If so, then it could obviously be affected in some way by the environment. And if *that* turned out to be true, depression could be a topic for epidemiology.

In 1971, the summer before Weissman was to begin her graduate work at Yale, she was hired to work as an assistant to Yale psychiatrist Gerald Klerman, who was interested in measuring the efficacy of two treatments for depression: a standard regimen of antidepressant drugs for the physiological symptoms, and a counseling technique he was developing, known as interpersonal psychotherapy, to improve everyday social functioning. His subjects happened to be forty women who had come to the Yale–New Haven Clinic for treatment of depression. Klerman assigned Weissman the job of coming up with some kind of test that measured how well these women functioned in their daily lives.

"I was a complete novice, so I looked up what was out there," Weissman said—anything to give her an idea of how one went about devising such a test and wording the questions. She found next to nothing. On one end of the scale, there were some "social functioning" tests that had to do with adolescents and courtship rituals; on the other end, there were tests given to patients in mental institutions designed to measure how well they could perform basic tasks of personal hygiene. It had never occurred to anyone to measure the social functioning of housewives. Housewives didn't "do" anything; what was there to measure? At the time, Weissman was in her late twenties, married, with four children under the age of seven. Even with the help of a full-time live-in nanny, she said, "I knew the energy it took to care for kids and do a decent job. This wasn't rocket science. It was common sense." She understood that these "nonworking" wives and mothers were performing jobs crucial to the functioning of their families, even if that work was not measured on any economic scale, and that depression could have a devastating effect on what they did. All she had to do was to come up with the right questions in order to quantify that impairment.

The scale she eventually came up with, in collaboration with another young psychiatrist on the project named Eugene Paykel, is still in use. It covered every major aspect of her subjects' lives, probing them for details nobody had ever asked about before. The questions ranged from the intimate ("About how frequently have you had sexual intercourse in the last two months?") to the kind designed to elicit information about day-to-day functioning: "Were there days in the last two months when you didn't do any housework? Could you tell me what those days were like? What have your feelings been toward the children during the last two months? Have you felt affection for them?"—and, breaking an ancient taboo—"Did you dislike them?" What she found at every turn was poor bonding between depressed mothers and their children and a pervasive irritability—"things you do not because you're bad, but because you're ill."

The first results of Weissman's research resulted in a paper that was published even before she started graduate school—"The Depressed Woman as Mother." That was followed in 1974 by a book coauthored with Paykel (now at Cambridge University), with the dismal but groundbreaking title *The Depressed Woman: A Study of Social Relationships*. By then, Weissman knew that she wanted to apply epidemiological research techniques to the study of how depression passes from one generation to the next. What were the long-term consequences of growing up with a depressed parent? What kinds of external factors could protect children from those effects? What kind of psychological attributes, such as emotional resiliency, also shielded children from the effects? What kind of treatment could be developed to intervene in parent-child relationships before long-term damage was done?

Physicians had noted for centuries that women suffered far more frequently than men from melancholy (or hysteria, or neurasthenia, or whatever name happened to be in fashion at the time), so the medical community greeted her research with great interest. It was a splashy start for a mere graduate student; she was breaking important new ground, with no competitors in view. "It amazed me," she said, "because I was pretty much a novice." Even the mainstream press picked up on what she was doing. "If anything, the ladies' magazines were all over it." The only real resistance she met came from her implicit assumption that depression in mothers would have urgent short-term and long-term psychiatric effects on her children. Standard psychoanalytic theory held that depression could only be a function of a fully developed ego; children did not have a fully developed ego; therefore, small children could not possibly be affected by their mother's depression, much less suffer from depression themselves.

The next phase of Weissman's research, then, was to see if her assumption was correct. In 1982, she began a twenty-year longitudinal study of ninety-one families, roughly two-thirds of whom had at least one parent who suffered from major depression. The ten-year outcome of that study documented that the children of depressed parents showed a roughly threefold increase in vulnerability to a long list of psychiatric disturbances: anxiety disorders, major depression, alcohol and substance abuse, various kinds of antisocial behaviors and social impairments, panic disorders, and phobias. (The twenty-year follow-up, published in January 2005, showed even more striking patterns. More about that later.)

Epidemiological research is crucial to the beginning of a scientific inquiry, because it defines the question; until that happens, everything else is anecdotal evidence. Despite the hundreds of academic papers she has published in the last twenty years, simply defining the question may be Weissman's most enduring contribution. At a time when depression was still seen as the product of amorphous mental "conflicts" rooted in the patient's childhood, she set about finding the answer to a more pressing and infinitely more practical question: "How does depression in the mother affect her children?"

As it happened, Weissman began her work documenting the generational legacy of depression at roughly the same time that technological developments were about to give neuroscientists something they had always dreamed of: a way to get a detailed look at the brain without destroying it in the process.

Before the 1970s, research into brain physiology was limited to actually taking a scalpel and digging around in the heads of dead people or—more grotesquely—lobotomy experiments in which doctors inserted the surgical equivalent of an ice pick between the eyeball and bony eye socket of a living patient in order to sever neural connections in the frontal lobes. (The technique didn't cure anything, but it could make agitated or violent mental patients very, very calm.) The only other way of getting "inside" the brain was via X-ray, which produced only blurry images of soft tissue—enough to see something that doesn't belong, like a tumor, but not much else.

But one of the marvels of scientific research—"the pleasure of finding things out," as physicist Richard Feynman memorably put it—is that discoveries in one area can yield unforeseen windfalls in an unrelated field. The work that laid the foundation for the techniques we use today to see the brain began with, of all things, pre–World War II atomic research.

By the late 1920s, scientists knew that the various parts of the atom—protons, electrons, and neutrons—spun around an axis, each with its own unique "wobble," and that each had specific magnetic properties. In the 1930s, Isidor Isaac Rabi of Columbia University discovered that exposing an atom to an external electromagnetic field, like a radio wave, made it possible to measure the atom's rate of spin and its wobble, which in turn allowed researchers to figure out the movements of the various parts of atoms in relation to one another. The technique was called magnetic resonance, and its discovery won Rabi the 1944 Nobel Prize in Physics.

From the beginning, there was obvious potential for using this tool in medical research: one of the simplest atoms to study is hydrogen, which is abundantly dispersed throughout the human body in the form of H_2O, or water. But the medical application of magnetic resonance had to wait for refinements in techniques and advances in computer technology. That happened in the 1970s, when British electronics engineer Godfrey Hounsfield built a machine that combined X-rays with high-speed computerized reconstructions to produce a kind of cutaway view of the interior of the human body—the technique we know today as computerized tomography, or CT scan. At about the same time, American and British scientists, working independently, were coming up with a way to refine magnetic resonance techniques so that it would yield information about not just the spin and wobble of a particular atom, but its physical location in relation to other atoms—in effect, a three-dimensional image. Then, using the same high-speed computer analysis methods that created CT scans, that information was used to create a far more detailed picture than a CT scan could produce. The name of this new technique: magnetic resonance imaging, or MRI.

A further refinement came with the development of functional MRIs (fMRI), which use contrasting agents injected into the body and track their path to essentially turn a series of still photos into a moving picture that allows researchers to watch the brain in action. By the early 1990s, MRI, fMRI, and positron-emission tomography (or PET scan)—a technique similar to fMRIs, which involves tracking the gamma rays emitted by a quickly decaying radioactive material injected into a patient's body—gave researchers a noninvasive way to get detailed pictures of the brain at work. For the first time, neuroscientists had a really clear view of an awesomely complex universe.

Today, one of the leading experts in the field of neuroimaging is a man named Wayne Drevets, who works from a tiny but extremely tidy office in an old house that is part of the National Institutes of Health in Bethesda, Maryland. The house sits atop the highest hill on the three-hundred-acre campus, and seventy-five years ago the original owner of the property would have been able to stand on the flagstone terrace and enjoy a lovely panorama of woods and pastures. Today, the view consists of a disorganized collection of buildings of varying vintages and ugliness, interspersed with insanely crowded parking lots. It was a hot day in May when we met, and the seventeen-year cicadas had emerged a week or so before. By the time I had finally found a parking space and made the long haul up the hill, I was sweaty and disheveled from swatting big black bugs out of my hair. Drevets turned out to be a tall, slim man with hair graying in a distinguished sort of way. He was wearing gray pants and a starched blue Oxford shirt and, naturally, looked as if he had never perspired in his life. It was hard to hold that against him, though, because he turned out to be a friendly and patient teacher.

A basic problem faced by researchers when delving into depression up to recently has been that there's no purely objective measure for it—no foreign body to see, no measurable substances to yield a definitive diagnosis. Even the most reliable depression screening questionnaires rely, in the end, on the patient's self-report. "It's not like diabetes, where you can measure blood sugar levels," Drevets said. The subjectivity in diagnoses has until now created all kind of problems: to take one obvious example, psychiatrists in Great Britain look at a set of symptoms and call it bipolar disorder, while their U.S. colleagues tend to diagnose the same symptoms as schizophrenia. Without empirical data, diagnosing a mental disorder is as much art as science.

Now, though, neuroimaging techniques can show us images of the distinct physiological characteristics of various illnesses: dementia, schizophrenia, bipolar disorder, depression. Eventually, Drevets said, neuroimaging will be able to discern the differences between various subtypes of depression. If nothing else, it's a tool that removes the last trace of doubt among those who might still argue that depression is "all in your head."

Which indeed it is, though not in any imaginary sense.

Drevets's work has shown, for instance, that there are subtle losses of brain volume among people with chronic depression in the area of the hippocampus, which has to do with the formation of long-term memory and the integration of memory with emotion. (By this point, I was thinking uneasily of an MRI of my own brain, taken the previous summer as part of a workup on a problem with migraines, in which my Russian neurologist had commented casually, "You have very odd-shape hippocampus.")

To illustrate, Drevets laid a page on the table in front of me. On it were comparisons of four images of the same slice of the brain: one from an anatomical dissection, one from a CT scan, one from an MRI, and one from an fMRI. The clearest image, far more detailed than the picture of an actual anatomical dissection, was the image taken with fMRI. The next picture was an MRI of the brain of an elderly person with major depression. The image showed a large, roughly X-shaped white patch deep in the brain where cerebrospinal fluid had seeped in to take up space created by dead brain cells—part of an increasing body of evidence that may someday explain the link between cerebrovascular disease and depression in the elderly (a distinct type of depression that occurs in people who may never have suffered depression before, and which is therefore different from the kind of chronic depression that develops, say, in adolescence—and, in turn, may be quite different from postpartum depression).

Next came two PET scans, comparing glucose metabolism in the brain of a healthy subject and in the brain of a subject with bipolar disorder. There were striking differences: the healthy subject's brain showed discrete and symmetrical bright areas of activity, while the areas of activity in the bipolar brain were a duller color, indicating reduced activity that was more broadly scattered and asymmetrical.

But what, exactly, do the pictures mean? Do they show the cause of an illness, or its effects? Nobody knows. But we don't have to know in order to benefit from spin-offs of such new knowledge. The discovery of the shrinking hippocampus, for example, recently led researchers at Columbia University to focus on the effects of antidepressants on that portion of the brain. What they discovered was that antidepressants stimulate new nerve growth there—a finding that, if replicated, would answer the nagging question of why drugs that get into the bloodstream within an hour can often take weeks to work.

In a crude sense, neuroimaging is like a flashlight: it helps researchers in a dark place know where to look. From advances in genetic research, for instance, we know that certain genes are found more abundantly in some parts of the brain than others. With neuroimaging techniques, we can now see which parts of the brains are active during various emotional states—fear, anxiety, depression, or mania, for instance. That, in turn, tells us which genes to look at in exploring the biochemistry of those emotional states.

Drevets believes that eventually neuroimaging can be used as a screening tool, to identify physiological markers associated with the illness in children who come from high-risk families. Treatment could then start early—even before the onset of any symptoms.

"This is one of the real hopes," he said simply, and I pictured my daughters, who are obviously at high risk. Then I resumed worrying about my very "odd-shape" hippocampus.

Like most mothers, my worries about my health center around my children. I have to be around, and functional, for—how long? Sometimes I figure it out. Until Suzanne is at least twenty-five? (No, make it thirty. Better yet, forty.) Because I deal with depression, some days present a bigger challenge than others—a fact that my children don't care much about now, but which I hope someday they will understand. At some level, I think they do already. Children instinctively know what effort is; after all, they've been engaged in hard physical and mental work since the day they learned to roll over. And, as I sometimes have to tell myself, if being a good parent is more work sometimes than at others, that's a reason to keep trying, not to quit. Why? Because the quality of mothering a child receives matters. We know this intuitively when it comes to human children. Now, from monkey children, we are learning exactly how.

The rhesus monkey mother squats on the ground, her long arms securely wrapped around her baby. The baby looks at the camera, wide-eyed, with an expression of serene curiosity. She is seeing a strange thing, but in Mommy's arms, she is not afraid.

Then Emory University neurobiologist Paul Plotsky clicked on another file, and on his computer screen came another short film clip. In this one, a baby rhesus monkey approaches her mother, who is sitting on the ground. The baby wants to suckle, and latches on to a nipple as best she can. Her mother continues to sit, arms by her sides—not resisting the baby's advances but not embracing her either. The suckling infant clings to her mother's chest the way she would cling to the trunk of a tree.

"Sometimes," Plotsky said, "you'll see babies of mothers like these just hanging around by Mom, just touching her toe or hanging on to the hair on her back. Anything to connect." He clicked on another image file in his computer, and there was footage of a baby approaching another female. "Watch this," Plotsky said. "The mother doesn't want anything to do with her baby but she doesn't want anybody else touching the baby either." Just as the other female monkey picked up the baby, the baby's mother came bounding over, snatching her baby away. For an excruciating moment, the two female monkeys fought over the bewildered infant, stretching its arms to an impossible length. Finally the second monkey let go, and the mother scurried away, clutching her baby under one arm.

"That may be a good football hold," Plotsky said, "but it is *not* the way to carry a baby. These are the baby monkeys who tend to wind up in the hospital."

He switched to another file. A mother monkey was sitting on the ground; next to her was her infant, and on the other side of the infant sat the infant's big sister. The big sister suddenly snatched the baby and began moving toward the compound wall, dragging the baby as if it were a rag doll. Then the older sibling settled down in a spot next to the wall and drew the baby monkey to her, giving the baby a halfhearted caress. Throughout, the baby monkey's mother remained indifferent.

"Wow," I said. "This is hard to watch."

"I know," Plotsky said. "There are some files I've deleted from my computer—pictures showing outright abuse— because I can't stand to watch them anymore."

We were sitting in Plotsky's office at Emory University; the videos were taken by some of Plotsky's colleagues at Emory's Yerkes Primate Center, about thirty miles north of where we were. At Yerkes, approximately 3,400 primates of various species live in a 117-acre compound that—except for the walls surrounding it—roughly approximates conditions they would encounter in the wild. The maternal behavior recorded on the video, too, shows the sort of thing that also happens in other monkey colonies and has been seen in the wild, Plotsky said, especially on the part of first-time monkey mothers. It turns out that the so-called maternal instinct is about as instinctive in monkeys as it is in humans, which is to say that it is hardly foolproof.

Monkeys are also remarkably like humans in other ways: when they are subjected to chronic stress, they become socially withdrawn, sleep too much or too little, act anxious and irritable—even drink too much—and, in general, show many of the signs that in humans we call depression. Genetically, too, there is a surprisingly high degree of similarity between humans and monkeys. But with monkeys, it's possible to do something you can't do with humans: manipulate the environment in ways designed to isolate the effects of parenting practices on behavior.

And here we come to one of the more intriguing discoveries in recent years: a study published in July 2003, in which Avshalom Caspi at the University of Wisconsin and Terrie Moffitt at King's College London reported that people who inherited a "short" version of a particular serotonin transporter gene were more than two and a half times as likely to develop major depression as people with the "long" version of the same gene. Both genes make the protein that recycles serotonin after it has been fired into a nerve synapse, but the "long" gene makes more protein and so works more efficiently.

Rhesus monkeys have the same gene and exhibit the same variation. In an investigation conducted at the National Institutes of Health primate colony in Poolesville, Maryland, NIH researcher Stephen J. Suomi discovered that monkeys with the depression-prone version of the serotonin transporter gene grew up to be perfectly normal monkeys if as infants they formed secure attachments to foster mothers who did not have the faulty gene. But monkeys with the depression-prone gene who were "peer-raised"—the primate colony equivalent of growing up in an orphanage—tended as adults to display most or all of the depressive behaviors listed above. In other words, Suomi concluded, "genetics and early experience factors can interact, often in dramatic fashion." What's more, "because [primate] daughters tend to develop the same type of attachment relationships with their own offspring that they experienced with their mothers early in life, such early experiences provide a possible nongenetic mechanism for transmitting these patterns to subsequent generations."

The nongenetic mechanism Suomi refers to is gene expression—the degree to which a particular gene is activated. Just having a particular gene in your DNA doesn't necessarily mean a thing; every person carries a complete copy of their DNA in every cell of their body, but only a tiny percentage of the genes in each cell get expressed, as geneticists say, which is what enables specialization—so hair cells can devote all their energy to growing hair, for instance, and not producing insulin. When it comes to disease, simply having a particular gene doesn't necessarily mean anything, either. You can have the gene for cystic fibrosis and never be sick in your life—although if you happen to produce a child with someone who also has the gene, chances are three out of four that your child will have the disease.

Genetically speaking, cystic fibrosis is a "simple"—that is, one-gene—disease. Depression involves dozens, maybe hundreds of genes, as well as the varying degrees to which those genes are expressed. And that, in turn, depends to a great extent on environment. What Suomi's research showed was that—in monkeys, anyway—a baby can be born with a gene that increases vulnerability to depression, but if that infant is raised in a caring and secure environment, that gene is not likely to be "turned on"—and the monkey will be just as resilient to stress as a monkey who never had the depression-prone gene in the first place.

This may sound obvious—after all, Little Orphan Annie got a lot happier after she was adopted by Daddy Warbucks—but deconstructing common sense on the molecular level is a complex undertaking. For one thing, until recently we haven't known which genes to look at (and the Caspi research focuses on only one; there are a number of genes associated with depression that come in more than one variation). Also, what is the biological mechanism by which this happens?

One possible answer is via hormones. In research on rats done by Michael Meaney of McGill University, Meaney found that the tactile stimulation provided by rat mothers who were highly attentive to their pups increases the levels of such feel-good hormones as oxytocin (which promotes bonding behavior) and dopamine (which is associated with the brain's "pleasure" system). That, in turn, "turns on" the genes that create receptors for those hormones, an effect that serves in the long run to mute the brain's response to stress. Rat pups born to mothers who displayed relatively little interest in their pups normally go on to mimic that indifferent maternal behavior when they have their own pups—except, Meaney found, when they were fostered as pups by the highly attentive mothers. If that happened, they became highly attentive mothers, too, and did not display any of the less desirable behaviors associated with their biological mothers.

Plotsky is collaborating with Meaney, and his current research focuses on the effects of stress in rat behavior—observing the behavior, measuring the effects on specific areas of the brain, and connecting those changes with specific genes. After we talked, he led me down into the labyrinthine maze of underground tunnels where he and his graduate students do their research. We took the elevator to the basement and walked down a long, meandering hallway. Then we came to a locked door, where Plotsky passed his identification badge across an infrared beam, and entered another long, dimly lit hallway, this one lined with gray steel doors. The air was thick with the smell of animal droppings and wood chips and whatever it is they put in rat chow. We put on surgical gowns and Plotsky put on a mask as well; over the years, he has developed a severe allergy to rats, and he rarely ventures into the lab without loading up on antihistamines.

"Here they are," he said, and opened a pair of steel doors to reveal a closet containing a large rack of rats—two per clear plastic bin, about six bins in a row, and about five rows to a rack. It was "daytime" in the closet, which meant these nocturnal animals should have been sound asleep. But as Plotsky opened the door, there was a collective shudder as the rats nervously shifted in their cages. They were black and white, about seven inches long minus the tail, with paws of an almost transparent pink. Their reaction to Plotsky's opening of the door was as if a siren had suddenly gone off.

Which sometimes it does. Being blasted at random intervals is just one kind of stress rats are subjected to. It's not the grisly physical torture animal rights activists describe, but if you were a rat it would not be fun: being startled awake by air puffs, being separated from your pups for indeterminate periods; being awakened by flashing lights. Many of these rats would be euthanized on the operating table when, after a massive dose of Nembutal, their brains are flushed with cold saline (a way of preserving brain tissue). I've always been in favor of killing rats, frankly, but now I couldn't help feeling disturbed. The kind of research that was revealing the mysteries of an illness I'd dealt with for most of my life inevitably entailed the suffering and death of living creatures, a fact I had conveniently managed to overlook. Now I was looking.

So, I asked, is this way of transmitting traits the same as saying depression can be inherited by ways other than genes?

"I wouldn't say 'inheriting,' " Plotsky said. "That gets people's hackles up, and they start thinking about Lamarckian theories of evolution"—referring to the long-discredited theory by the eighteenth-century botanist Jean-Baptiste Lamarck, who believed that traits acquired by an organism during its lifetime could be passed along to the next generation. To put it more accurately, Plotsky would say that being exposed to particular maternal behaviors (neglect, for instance) alters gene expression in the mother's offspring. That, in turn, is likely to produce the same kind of maternal behavior when that offspring grows up to have pups of her own, and so on down the line. But it's an effect that can be reversed: even rat pups born to neglectful mothers can be put in environments in which the genes for undesirable behaviors become muted, and otherwise dormant genes get turned on. Nothing about the basic DNA of either group of rat pups changed; what changed was the level at which various genes got activated. Environment alters genetic expression. Genes influence, but do not dictate.

The potential applications of all this are enormous—if these discoveries are applicable to humans. So far, the evidence is that many of them are. Studies by Tiffany Field, of the University of Miami, have shown that the fetuses of mothers who are depressed move around more, and that after birth those babies display a "profile of dysregulation" for the first year of life: inconsolable crying, abnormally high levels of stress hormones, and disorganized sleeping patterns, among other things. Research by Field and others has shown that these infants show elevated levels of cortisol, the stress hormone, for well past the first year of life. Emory University neuroscientist Charles Nemeroff has theorized that early-life stress—sexual or physical abuse, in the context of his study—permanently increases the brain's sensitivity to stress. Field's research suggests that those changes begin even before birth. More recent studies have shown that the hippocampus in women who suffered physical or sexual abuse as children was smaller than normal. Clearly, the brain bears the scars of psychic stress, just as other parts of the body bear the scars of physical trauma.

"At this late stage in my career, I know I'm not going to find a cause or cure of any disease," Plotsky said. "But one thing will remain true in my kids' lives, and that's that there will always be more need for treatments for mental illness than there will be services available. If we can understand these processes and where these sensitive windows of development occur, we can target our interventions"—perhaps someday figuring out therapies or drugs, or both, that can be used with children who are at high risk for depression during crucial points in their development. "We can create interventions that produce resilience."

It's already clear that the chain of transmission can be broken. In January 2005, Weissman published her most recent research on the ninety-one families she began following in 1982, who have now produced grandchildren. Of the 161 children, those whose parents *and* grandparents had suffered from moderate to severe depression, an astounding 67.6 percent already had acquired some kind of psychiatric diagnosis by the time they were at or approaching puberty. But then there was a surprise: in families where one or more parents in the first generation had suffered depression, but the second generation had not, the percentage of grandchildren with a psychiatric diagnosis fell sharply, to 31.3 percent. In short, the longer depression snakes through the family tree, the more likely it is to be passed along. But alter the environment—make it a situation in which a child is exposed sporadically to grandparents who are depressed, but where the parents act as a buffer—and the pattern changes. In theory, then, there seems no reason not to assume that *any* lessening of exposure—say, early and effective treatment for a depressed parent—would not produce a similar benefit. Human resilience is like a weed: give it half a chance, and it will grow.

cᴑℴ

So, at least in theory, environment can affect genes. Can genes affect environment? That is, can your genetic makeup play a role in things like what kind of people you choose for friends, what types of places you relax in, whether you ever become a crime victim?

Kenneth Kendler of Virginia Commonwealth University in Richmond thinks the answer is yes. He has spent the last twenty-five years or so studying several thousand people whose names were culled from something called the Virginia Twins Registry, a database of all the twins born in that state after 1918. Kendler and his colleagues tracked down as many of these people as possible and asked them to participate in the study. They ended up with more than seventy-five hundred participants—identical and fraternal twins of both genders. In some pairs, both twins suffered from depression; in others, only one twin did; in others, neither twin had a history of depression. The aim, Kendler said, was to get a population sample as representative of the general population as possible.

Kendler then came up with nine basic categories of stressful life events: being an assault victim, having serious marital problems, going through a divorce or breakup of relationship, job loss, losing a close friendship, serious illness, major financial problems, being robbed, having serious legal problems, death of a close friend or relative, illness of a close friend or relative, and having a major argument with a close friend or relative. When he looked at identical twin pairs in which one twin had a history of major depression, he found that the other twin had a significantly increased risk of encountering six of those nine categories of stressful experiences. In non-identical twins where one twin had a history of depression, he saw the same effect, but in a more muted form. In short, the twins with two copies of the genes predisposing to depression met with significantly more problems in life than those with only one copy—and those, in turn, had more problems in life than twins without any genetic risk.

This wasn't just bad karma, he concluded. "Genes for psychiatric illness may also impact on risk by causing individuals to select themselves into high-risk environments"—tumultuous romances, arguments with the boss, irritability that causes chronic stress in a marriage.

A lot of research, including Kendler's, has found that people with a genetic vulnerability to depression tend to be people who score high on tests that measure "neuroticism"—psychology shorthand for traits we might call moodiness, or being anxiety-prone, easily stressed, irritable, or sensitive. These are traits that, in some contexts, might produce a lot of writers and artists (and, in fact, the rate of mood disorders in such groups is suspected to be much higher than in the general population) and which in other contexts just produce people who are hard to be married to or work with. No matter the context, these are people who are particularly sensitive to stress. Random life events also play a powerful role: women in Kendler's research who had a history of childhood sexual abuse *and* a genetic vulnerability to depression are particularly likely to develop the illness. But—most important—

"even in the presence of high genetic risk and severe stressful life events, the majority of individuals do *not* develop an episode of major depression."

Kendler's most recent paper deals with the effects of social support networks on the risk of depression in men versus women. In a survey of more than a thousand male-female twin pairs, he found that among women, low levels of social support were strong risk factors for depression. Among the men, in his words, "it didn't matter one whit." The importance of a support network for women, and its lack of importance for men, was the most dramatic gender difference that he has ever turned up.

As Kendler spoke, I thought about a period of my own life that seemed to illustrate what he was talking about—a time when I was in my early thirties and had just moved from Atlanta to Washington, D.C. I was single and lonely; my twenties had been marked by episodes of depression and a series of failed relationships. Now I was living for the first time without the extensive girlfriend network I had built up in Atlanta. Then I met someone, a widower with small children. There was a strong and instant mutual attraction. Yet within a week of our first date, something happened that made it clear he was controlling, irrationally jealous, and could even be violent. My instincts said, *Stay away*. But all those years of intermittent depression had left me without much experience to back up my instinct, and so I flung myself into what quickly became a verbally abusive relationship. The three years it lasted included the worst depressive episode of my life. I was hospitalized once, and twice came close to suicide.

It was an apt example, Kendler said. "You selected yourself

into this high-risk relationship, and you were also sensitive [to depression]. In abusive relationships, a fair number of people can survive them and not get depressed, but it's very likely that would push you over the edge into depression." My genes, in short, helped to create the environment in which my illness would flourish.

So, to summarize: your children may have inherited your "bad" genes—but depression is a lot more than just genes. In fact, the more complex the picture becomes, the more encouraging it gets. In 1982, what little published research there was on maternal depression had a distinct "blame the mom" tone reminiscent of the days when psychiatrists believed that emotionally remote "refrigerator mothers" caused schizophrenia in their children.

But as Kendler's research shows, we are not passive recipients of either our genes or what life hands out to us. This is poker, remember, not chess—and much depends on how we play the hand we are dealt. As a friend of mine who suffers from multiple sclerosis once said to me, "I have MS—MS doesn't have me." A diagnosis that would make some people sink into despair was, for her, a reason to change career paths: trained as a lawyer, she now runs a nonprofit advocacy group for people with the illness. I don't know if I can match her abundant emotional resilience, but I can still see the difference personal choices have made in my own life. At a crucial point, when I was in the hospital, I was able to find some doctors who could effectively treat my depression with medication, and I chose to take advantage of that. Being hospitalized in itself was an important first step in giving me some distance from the abusive relationship I was in, and it marked the beginning of several years of psychic housecleaning. Much of that involved revamping my social support network. Slowly, my life began to get a lot better—not perfect, not always easy, but definitely better.

Throughout life, we get chance after chance to make crucial decisions: what kind of person we marry; what kind of work we do; the kinds of friends we choose; whether we slide into intellectual or spiritual stagnation, or continue to learn—and what we teach our children. As Kendler notes, people often make the mistake of equating "genetic" with something that is unchangeable, and "environment" with something that can always be manipulated. Sometimes the opposite is true. A genetic vulnerability to depression can be influenced for the better by choosing to get medical treatment; something purely environmental—say, prenatal exposure to a carcinogen—can result in irreversible damage.

If your children have inherited a vulnerability to depression, it's no catastrophe. As wise old Professor Albus Dumbledore once told Harry Potter: "It is not our talents [or genes] that define us, it is our *choices*."

CHAPTER SIX

Don't Look Now, but Your Kids Are Stealing Your Coping Skills

Whether my depression has been a benefit or a hindrance is up to [my son, and] how he chooses to see it.

—ANONYMOUS SURVEY RESPONDENT

There are two basic ways in which depression is handed down from one generation to another: by the genes parents pass on to their children, and by the behaviors parents model for their children. These days, genetics is the all-absorbing subject, which is not surprising. But suppose you knew for a fact that you possessed the short allele of the 5-HTT serotonin transporter gene, and that this explained why you have always been vulnerable to depression. Suppose also that you knew you have passed along this gene to your children. On a day-to-day basis, how does this knowledge change your life? Not much.

On the other hand, how you respond to your own depression matters immensely. The ways that you deal with stress, the assumptions you make about life (both the conscious ones and, more important, the ones you've never really thought about), the ways you handle adversity, how you express anger, the responsibility you take for your own health, how you recover when you make mistakes—the sum of all of this offers your children their first and most important lessons about how the world works and how they might fit into it. The only thing more certain than this fact is that, sooner or later, you will screw something up and become a horrible example. "They fuck you up, your mum and dad. / They may not mean to, but they do. / They fill you with the faults they had / And add some extra just for you," wrote Philip Larkin in another of his frequently quoted poems. When I first ran across that poem in my twenties, I felt a prim sense of vindication, and its sour admonition ("Don't have any kids yourself") seemed an attractive option. Now, having had kids, it just makes me laugh. Anyone who has children knows, or should, that along with immense rewards, some failure is inevitable. Keeping this in mind leaves you free to do the best you can— which, in any event, is all you can do.

Depression, however, adds another layer of complexity to the task of nurture, by subtly warping our perspective of our own place in the world. The emotions of depression are intense; just as intense is the innate human drive to make sense of what we feel. The result, too often, is that anyone who struggles with depression makes assumptions about reality that are based, at best, on half-truths. It is a difficult and counterintuitive exercise to say, in the midst of a depressive episode, "These are just emotions; they don't have anything to do with reality." It's much easier to pick out the facts that support your feelings. When you feel like a failure, for instance, your mind edits out all the times you succeeded despite the odds. All you remember are the other times you failed. Your mind says, *See? It's always been this way. It always will be this way.* And so, feeling like a failure, and justifying it with what seems like abundant corroborating evidence, you behave in ways that teach your children that passivity is the safest course to take, that risk-taking never pays off, that the odds are stacked against them. The "reflector effect" happens in other ways, too. A mother who thinks of herself as ugly raises children who may hate what they see in the mirror; a mother who doesn't find much to like about herself is likely to raise children who will wonder if there's anything worthwhile about themselves. A mother who interprets every personal setback as a permanent catastrophe has nothing to teach her children about resilience or perseverance.

This happens all the time, despite immense love and the best of intentions. By the time a child is old enough to fully articulate her self-doubt, it will be hard for any chronically depressed mother to muster reassurance that does not ring hollow. When I was tortured by self-loathing as a teenager, my mother would tell me how smart and pretty I was. Most mothers of teenage daughters spend time reassuring their daughters this way—it goes with the territory—but by this point I had spent years listening to my mother disparage herself. There was no way she could help me. *How would* she *know about anything?* I would think despairingly. *Hasn't she been telling us for years what a dimwit she is?*

These habits of thinking that parents pass along to their children are "explanatory styles," in a phrase coined by University of Pennsylvania psychiatrist Martin Seligman, and they are every bit as heritable as blue eyes or singing talent. The heritability of nongenetic influences is a complex concept, but it's widely and intuitively understood. As one mother in our survey wrote, "I think (depressive) thoughts are passed on and on and on, generation to generation, getting worse each time"—a way of putting it that sums up the idea and, at the same time, neatly demonstrates depressive thinking at work.

The most interesting thing about explanatory styles, however, is not their heritability, but their changeability. You may inherit a certain way of looking at the world from your parents, but at any point in your life you may choose to reexamine your core assumptions and—just possibly—alter them. Choosing our core beliefs and assumptions about life is a momentous undertaking, and nobody gets through life without doing it; not to decide is to decide. Much in life is out of our control, but how we interpret events is largely a choice we make for ourselves. In fact, it may be the *only* thing we control. How much difference does this choice make? It's a question with no clear answer. But in recent years I've begun to see how some of those choices have played out in my mother's life. I've also glimpsed tantalizing hints of how things might have been different—and how, with any luck, they may still be.

"Well, I just about fainted!" my aunt Dede writes. "Ted brought the letter to the store last Friday morning 'cause he goes to work at twelve, and I screamed when I read it. I cried and laughed and just poured sweat. Everyone thought I had a miscarriage. I never laughed so hard in all my life."

I had found the letter in a bulging manila envelope in my sister's basement, among the truckloads of stuff we had carried out of my mother's house after her stroke. Inside were forty or fifty letters from my mother's sister, covering roughly the decade between 1953 and 1963—notes scribbled during the baby's nap, after the laundry was on the line, late at night after everyone else was in bed, during little niches of solitude carved out in the predawn hours. Among them were scattered clues to a question that, ever since my children had been born, I had grown intensely curious about: What had motherhood been like for my mother?

Long ago, my mother had erected a wall around her emotional life; I had always known that her most important feelings (with the exception of her thoughts on my moral failings, of course) were things I would never hear from her. And then came her stroke, which played a cruel trick: it cracked the wall—she had never talked about suffering postpartum depression (PPD) before then—but it also erased many of the memories inside. *Did you ever feel like . . . ?* I wanted to ask. *Were there days when you just . . . ?*—and, more and more, she would say, "Oh, honey, I don't know. You know how my memory is." As generations of women before me had found, there is something about becoming a mother that makes you miss your own mother; it is a profound and primal yearning. So when I found that manila envelope, I read the letters avidly. I was a biographer looking for clues, a daughter hungry for at least an echo of her mother's voice.

In 1953, my mother and father had just moved into their first real home, just south of Atlanta. Dede was in Los Angeles, pregnant with her second child. The date of the letter told me that my mother had written with the news that she, too, was pregnant—for the first time—and that their due dates would be almost identical. Dede was ecstatic. "I want you to be a mother the same time as me," she wrote.

Dede and my mother were half sisters, four years apart. Dede's father died in 1924, leaving behind his Masonic apron and not much else. My grandmother Hulett Ruth Horton supported herself and Dede by working at the Nabisco factory in Atlanta, stapling cracker boxes together for $9 a week. My aunt recalls their mother, who went by her middle name of Ruth, as emotionally reserved and devoutly religious, with beautiful long, dark hair. "I don't remember her kissing either one of us," my aunt said. "I don't remember laughing or fun or games."

Not long after her first husband's death, Ruth Horton met a tall, dark-haired man from Greens Creek, North Carolina, named Enley Eugene Buchanan. The handsome visitor who came to court my grandmother made a big impression on her little girl by presenting her with a box of chocolate-covered cherries. They were married, my aunt tells me, by a justice of the peace at a courthouse somewhere in Atlanta, although I've never been able to find a record of this. In 1926 my mother, Enley Ruth, was born. After her birth, her father seemed to fade from the scene, a circumstance partly explained by the fact that he worked on major bridge and tunnel construction projects throughout the Southeast. But my research also turned up an assertion that about the same time my mother was born, he was involved in a relationship with a woman in Augusta, Georgia, and fathered a daughter by her. My grandfather, it seems, suffered from wanderlust and what we might today call commitment phobia. He would show up from time to time to bring money, my aunt recalled; each time he left, their mother "just cried and cried."

The story my aunt tells of their mother's death sounds like a mawkish 1930s melodrama: there had been "a beautiful gray coat with two rows of black buttons" belonging to my grandmother, which she cut up to make coats for her daughters in the fall of 1931 when she had no money to buy them coats. It had been a typical Atlanta autumn, with temperatures in the sixties, but in the first week of November the temperature plunged into the twenties. My grandmother was wearing a sweater to work, and had evidently been nursing a cold or mild case of the flu. In the sudden chill, it became a deep, bone-rattling cough. On November 5, she was taken by ambulance to Grady, Atlanta's charity hospital, suffering from pneumonia—a serious complication in those pre-penicillin days. A neighbor drove my aunt and my mother to the hospital to see their mother. "She waved at us from a stretcher going into her room," my aunt said, "and that was the last I saw of her." On her deathbed, my grandmother entrusted her two little girls to one of her three brothers. My mother was five; Dede was nine.

But the Hortons were poor, and it was the Depression. Dede and my mother bounced around among their uncles, overhearing heated conversations about who should get saddled with these extra two mouths to feed. At one uncle's home, they were fed rancid scraps from the school cafeteria where their uncle's wife worked. When social workers found out about that, my mother and her sister were placed in the Southern Christian Children's Home in Atlanta. Throughout,

Dede was the closest thing my mother had to a maternal figure. Once, after losing a fight with some other girls, my mother remembers crying on the playground and feeling Dede's arms enfold her. "Poor little kid," Dede said. "I'll be your mommy."

My mother was adopted when she was eight by a kindly Southern Railway worker and his wife. They could afford only one child, but they made sure Dede and my mother saw each other frequently. Dede left the orphanage for a series of unhappy foster homes, married at sixteen, moved to California, had a daughter, got divorced, and then married the nice man I remember as Uncle Ted, a manager for a Ralph's grocery store. At the same time, my mother was growing up to be the red-haired beauty who caught my father's eye one day in 1946 in the reservations department of Delta Air Lines. In September 1953, both Dede and my mother gave birth to girls—my cousin Candy and my sister, Ellen. Motherhood became their shared avocation, and for the next decade the letters from Los Angeles were filled with the minutiae of raising kids: growth spurts, childhood illnesses, teacher conferences.

My aunt did not keep my mother's letters, and there is only one letter in the pile written by my mother—a typewritten carbon, dated November 28, 1955. I would have been three months old. "I have been looking for a letter from you for days," it begins. "Guess you don't love us anymore, huh? Or is Candy sick?" (Candy had been born with a serious heart ailment.) My father was on the road again; he traveled more often than he was home. "I wish I could get a [Delta employee's] pass to come out to see you—we get awfully lonesome while he is gone," my mother's letter went on. "Not that

I don't have enough to do—but I just have to get away some-time or go nuts."

Three thousand miles away, Dede was lonely, too—at home without a car, married to a man who was married to his grocery store, preoccupied with the care of a frail baby whose physical ailments seemed unending. "I worry and cry all the time," she wrote. "I guess I have never faced a problem like this before and I sure do feel sorry for my little doll." As if that weren't enough, her third child, Teddy, was born with severe colic. "I get maybe three hours' sleep a night," she wrote when he was about two months old, "but it's all broken up so it doesn't do me much good. Sometimes I fall asleep eating my dinner." There is no self-pity in the tone. In those pre–*Feminine Mystique* years, women like my mother and Aunt Dede took exhaustion and loneliness for granted. A good husband was one who brought his paycheck home, and their husbands did.

"In your letter you said you had had your share of troubles for the last three or four months," my aunt writes sometime in 1958. "What's wrong?" (That's like my mother, to hint darkly about heavy burdens and then refuse to elaborate.) "You sounded real down in the dumps," the letter continued. "I sure do get that way but I was hoping you didn't. I guess everybody does sometimes, huh? I had myself a good cry last night 'cause poor Candy came in with one of those crazy stomachaches . . . and she was whimpering and I was rocking [Teddy] and had been for 1½ hours and I wanted to go to her and I couldn't 'cause his eyes were almost shut and I felt so darn sorry for her. I got so mad I just bawled."

Yet the personality that emerges from my aunt's letters is

cheerful. In one letter from the late 1950s, my aunt describes a three-week-long series of health crises and mentions a note from Candy's teacher that is "about the only words of encouragement we have ever had about her"; by then, it had become clear that Candy's heart condition had affected her mental development. But in the next paragraph, she launches into an ebullient description of her kitchen remodeling plan. "Jeepers, I can hardly wait!" Her letters also frequently mention neighbors and close friends; it was clear she had a support network.

My mother didn't. We had only a handful of neighbors, and my mother's early years had ingrained in her a fear of rejection that made it hard for her to cultivate friends. She had none of her sister's resilience. She loved my father's jokes, and I remember times when she could be silly and playful. But there were also days when she banged the dishes and jerked the vacuum cleaner around the living room as if it had personally offended her, when she barely spoke to us and disappeared behind her bedroom door for long periods to lie down for one of her blinding headaches. When I saw that vertical frown line between her eyebrows, tension would creep down my spine; after a while, I would become aware that I was holding my shoulders stiffly and that I had a headache, too. I worried about her; I wanted to make her happy. I brought every achievement home for her admiration; I told her jokes; I clowned around to make her laugh. But there was a darkness around her I could never fully penetrate.

Church was her refuge. The hellfire preaching I heard there exacerbated all my anxieties, which were already rampant, but my mother clung to the idea of a God who would punish the wicked and comfort the afflicted. In the next life, God would wipe away all tears, "and there shall be no more death, neither sorrow, nor crying, neither shall there be any more pain." It was a promise of both reconciliation and retribution; to the orphan who had known neglect and cruelty, it must have been deeply appealing.

So the years passed. As they did, the intimate tone of my aunt's letters began to fade, replaced by descriptions of unending rounds of home improvement: wall-to-wall carpeting, a glass-enclosed sunroom, a color TV to replace the black-and-white. Year after year, there were reasons why it was not a good year for us to visit, and, as time went by, stories of a lifestyle my mother found alien and un-Christian: a neighbor's divorce, drinking, gambling trips to Reno. There was one visit, in the summer of 1963, that remains vividly in my mind—the palm trees, the cold smell of air-conditioning (not really a smell but the absence of smell), the novelty of bottled water. My uncle kept beer in the refrigerator, and my mother's silent disapproval filled the house. I thought my aunt was glamorous—so slim in her capris, with her bleached-blond bouffant, gesturing with a cigarette, and telling stories in a voice that, even after decades in Los Angeles, had never lost its Southern drawl. Dede thought my father was hilarious. At night I drifted off to sleep in my cousin Candy's room upstairs, listening to bursts of laughter from below.

But it was the last visit. When my aunt and uncle divorced, there was an abrupt chill in the communications between Aunt Dede and my mother. Years later, I learned that my mother had written Dede a letter denouncing divorce as immoral. She must have been frightened that her sister was becoming a stranger; the harshness of her judgment is, I think, a testimony to her terror. Yet in retreating to the mountaintop of her religious beliefs, she managed to create the very estrangement she dreaded. It would not be the last; years later, she interpreted my abandonment of the fundamentalist church as cruel and personal repudiation, and I think she saw it coming before I did. There was a wistful tune I remember from my childhood, made famous by the Kingston Trio. "Where are you going, my little one, little one?" they would croon. "Where are you going, my baby, my own?" Whenever they sang it on the radio, the child my mother looked at was me.

My mother's story is painful, but here's the funny thing: all in all, she has had an amazingly lucky life. She was adopted by loving people. (My aunt never was.) She and my father were happily married for thirty-three years. (My aunt was divorced twice and widowed once.) Her children were healthy. (Of my aunt's three children, only one lives today as a self-sufficient adult; her oldest child, a daughter, died tragically a few years ago of complications from gastric bypass surgery.) My father's career took off in the late 1960s and we moved into a house on a hill with an eighteen-acre spread, a circular driveway, and a sunken living room; by the time he died at fifty-nine, of cancer, he left my mother a comfortable estate. (My aunt has never had any money to speak of, lost what little she had after her second divorce, and continued working until heart problems forced her to retire in her seventies.)

It took several years for my mother to come out of the shell of her grief, but slowly she did. She sold my dad's old Volvo, bought a peppy little car with a sunroof, and found a group of Delta retirees who liked to travel. For a while, in the mid-1980s, I told friends that I couldn't keep track of her: one week she was in Venice; the next week she was boating down the Danube. And then the real shocker: at a high school reunion, she reencountered the man she had dumped to marry my dad. His wife had recently died, and overnight, the romance between him and my mother bloomed anew. Even after her stroke, in 1995, he never left her side. Until the day she died, he visited her faithfully in the nursing home. In a world where many women despair of finding the love of even one good man, my mother found it twice.

And yet, in her eyes, happiness was an iffy proposition. Along with happy memories, her mind also contained a mental library of every scary thing she had ever heard, every disappointment she had ever suffered, every person who had ever hurt her—back to and including the little boy who pulled the chair out from under her in grammar school. Life held no guarantee of love; the only guarantee was that bodies would sicken and age and that people you love would leave. When my sister and I were small, she often reminded us that we were blessed to have the kind of happy childhood she had missed. "So enjoy yourselves now," she would add, "while you can!" If by chance we *were* having fun, that remark would always squelch it.

My aunt, in contrast, has rolled with the punches throughout a much rougher life. "I thought Ted would leave and I would stay," she told me once, speaking of her divorce back in the 1960s, "but oho!—the joke was on me. *He* kept the house." She chortled. When I asked her for recollections of my mother's father and their early life, she told me what she could, and then berated me for my morbid preoccupation with the past. "Your mother has had an ideal life," she added, "and I think it's sad that she [doesn't] realize how lucky she is." Well, "ideal" is stretching it, but my aunt does have a point.

My mother's worldview is a good example of what Seligman calls a depressive explanatory style, which he breaks down into three parts. First, people who think this way tend to see themselves as the primary cause of negative events. When I ran into a brick wall with algebra in high school, my mother would say, "Well, you get it from me, honey. I never could understand math, either." My problem wasn't really my problem; it was *her* failure to bequeath me a math gene. If she tried a recipe that didn't turn out, it never occurred to her that maybe combining rhubarb with chicken was an inherently flawed concept. No, the problem was *her*. It was a family joke that she never put a meal on the table without apologizing for it. "I don't know about this roast," she would begin dubiously, and my father would say, "Stand by—here it comes." (In fact, she was a good cook.)

Second, a depressive explanatory style is pervasive, which is a good way of describing my mother's worldview. She told me once that her childhood experiences "were the beginning of feelings of rejection that have stayed with me all my life"— and, in fact, rejection or abandonment (perceived or real) has been a theme of every close relationship she has ever had. Her real mother died; her father disappeared; her adoptive mother had not loved her (at least, that was how she remembered it); her adoptive father had died—on Mother's Day, no less; her sister/surrogate mother "rejected" her; her husband died young. Years later, she would enact the same anguished drama with her best friend, who had been as close to her as a sister, and, in different ways, with both her children. To her, abandonment, love, and loss were three strands of the same rope.

Finally, Seligman says, the depressive explanatory style is rigid. Bad events are permanent and unchangeable; nothing ever changes for the better. When I told my mother I was moving from Atlanta to Washington, D.C., she was horrified. It was well known, she said, that women outnumbered men in Washington about three to one; if I moved there, I would be throwing away any chance of meeting Mr. Right. Later, I found out the three-to-one statistic had stuck in her head from World War II days, when it may briefly have been true. (Washington, by the way, was where I met my husband.)

These examples may sound innocuous, tragic, or even funny—but the cumulative effect is insidious. Sons fall in love with mom at some point; daughters model themselves on her, at least in some ways, whether they intend to or not. In truth, my mother is a smart woman with natural leadership skills and a strong creative bent. But her self-image is utterly at odds with the evidence. Why a brilliant man like my father had been attracted to her was a mystery; the fact that her daughters turned out to be intelligent and attractive was due entirely to him. "I just hatched you," she often said.

And yet, despite her childhood traumas, even my mother showed signs of being able to change. In the fourteen years between my father's death and her own crippling stroke, she managed to summon up reserves of strength and resilience I never knew she had. In a way, facing her worst fear—life as a single woman, living alone for the first time—made her a different person, open to solitary adventures but also sociable and full of energy. To me, of all the tragedies in my mother's life, the worst was not that she never really got to be a child; it's that she had such a brief time to be an independent adult.

By then, of course, I was an adult myself, and through some mysterious kind of mother-child alchemy, I had long ago absorbed her pain. All children fear losing their parents, but not many spend night after night kneeling in anguished prayer, as I did, imploring God not to let their perfectly healthy mothers die. My mother passed down many things to me—everything from some of my most deeply rooted moral values to how to fold fitted sheets—but the first, and most contagious, was fear. She did not "give" me her depression, but the attitudes I learned provided fertile soil for it to grow.

And I, in turn, handed this legacy down to my older daughter.

Why was I so stunned when this happened? Why didn't I suspect it at the first sign of trouble? To begin with, Rebecca's major symptom was extreme irritability, which is a common sign of depression in children—but depression in children is very different from depression in adults, and I didn't know that. She had none of the other symptoms of depression I was so familiar with. But the main reason was plain old denial. This was my *little girl*. Surely, *seven*-year-olds do not get depressed. But they do.

Of course, to say I handed down my depression to Rebecca is a reductionist view. At most, I can take responsibility for only half of Rebecca's DNA and some of her nurture—

and, in fact, depression is a vine that grows on my husband's family tree, too, though not as luxuriantly as on mine. There's also the incontrovertible fact that Rebecca is her own self, that babies enter the world with their own unique personalities. But still: remember Tiffany Field's work on babies born to depressed mothers and the "profile of dysregulation" they exhibited? Rebecca and I fit the profile.

The summer I was pregnant with her, a combination of job and family crises upped my stress level sky-high: my mother had recently had her stroke, her finances were in disarray, I hated my job, I was having a lot of minor but extremely uncomfortable health problems associated with my pregnancy—and then, PPD struck. It is not coincidence, I think, that Rebecca was born with a brain that seemed to have more hot wires than a stolen car. For her entire first year, she did not take a nap that lasted longer than forty minutes. When she hit her terrible twos, her tantrums were remarkable for their ferocity, their duration, and for her rigid inability to be distracted or consoled. One in particular sticks in my mind—an outing to an aquarium when she was twenty-two months old. It started with a silly thing—she wanted to get into her stroller by herself, and couldn't—but she screamed so loudly and so long that the security guards threw us out of the building. She then put on a howling, hair-pulling, frightening sidewalk show for the passersby that lasted for almost an hour before I could finally wrestle her into in her car seat and slink home. The rest of the day was the same: short intervals of calm interrupted by hysterical rage. That night, I picked out a pair of pajamas for her. She screamed "NO!"—and it was the last straw. I sat down on the floor and began to sob. Rebecca stared at me. After a few moments I reached out to her and drew her close.

"I'm sorry, honey," I said. "Mommy is sad today."

"I sad, too," she whispered.

For the next five years, our lives would center around Rebecca's hyperirritability and anxiety (which was extreme—this was a child who stayed up until 2 A.M., watching for prowlers, who ran screaming to me every time the garbage truck showed up). I took parenting courses and made David take them, too. Yet the firmer the discipline we imposed, the more stringent the "logical consequences," the wilder she became. She shattered mirrors, she slammed doors, she destroyed her own toys. I read books on attention deficit disorder, "explosive" children, "spirited" children, bipolar disorder—she fit some part of all those descriptions, and yet nothing seemed exactly right. It wasn't until the end of a nightmarish year in second grade that we finally got Rebecca tested and—by flagrantly abusing my media connections at the National Institute of Mental Health—connected with a first-rate cognitive behavioral therapist, Dr. Ruth Stemberger. She had set to work with Rebecca by handing Rebecca a piece of paper and a marker. "Now, draw me a picture of your brain," she said. Rebecca had produced a remarkably sophisticated rendition, which Dr. Stemberger used to explain in simple terms the difference between the limbic brain and the cerebrum. "Now, what happens sometimes is that the 'mad' part of your brain takes over your 'thinking' brain," she said. "We have to come up with some ways for you to get back in control." It was a neutral, nonjudgmental way of explaining Rebecca's tantrums, and the emphasis on "taking control" was a shrewd way of enlisting Rebecca's cooperation. Dr. Stemberger in turn led us to Dr. Peter Dozier, a soft-spoken man with a kind face and sandy-gray hair who walked into his office waiting room, said a perfunctory hello to me and David, and then turned to Rebecca. "It's so good to meet you!" he said, with what seemed like genuine pleasure. He spent two hours with Rebecca, and then called us in to tell us his diagnosis. It was straightforward: childhood depression.

I felt shock, and a wave of sadness, and also a certain relief. I had known that something was wrong; I had felt it every day for the past five years like a physical weight in my chest. All the energy I had poured into getting the right help for Rebecca—a process that had taken the better part of two years—had been based on my assumption that I could ward off whatever was stalking her, that by being proactive I could prevent depression in her future. Yet all that time, depression had been quietly smoldering in her brain. When had that spark ignited? "I sad, too." Could a two-year-old possibly suffer from what the Greeks called "fear and sadness without cause"? Did the spark ignite in my womb?

Dr. Dozier was looking at me. "I think," he added carefully, "there has been considerable suffering." He did not say it accusingly, or pityingly, but as a validation: *You are not imagining this, your child is not a brat, you are not rotten parents*. Then, unexpectedly, he smiled. "Things *will* get better," he said. "You'll see; they will. I don't know of any family where a child has suffered like this that the family hasn't suffered also. But things *will* get better."

We left that day with a prescription for Lexapro, a tiny white pill that we painstakingly cut up every night into fourths and gave to Rebecca at dinner. In this mundane, daily act we were entering a brave new world in which the line between personality and disease was blurred, in which what once might have been called character defects were reduced to the abnormal behavior of molecules. This was scary stuff. Were we doing chemistry experiments with our daughter's developing brain, or helping to heal it? I'd read the stories about antidepressants and suicidal behavior in children; I knew the risks. But watching this little girl trying to learn from therapy while tossed about by her volatile moods was like watching a child trying to learn how to sail in the middle of a storm. She needed calm waters.

So we took the risk, and slowly things began to change. The child who once spent much of her time sullen with rage is now an exuberant eight-year-old who groans at my jokes and helps her little sister put on her pajamas (when she is not bossing her little sister around), who runs with coltish grace, who can sing the entire sound track of *Seussical*. She is interested in drama, a natural outlet for a child who is a born diva. She gives me fierce hugs and way too much attitude. The child who once could wreck a playdate in ten minutes now has *friends*.

One night not long ago, she came home from a daylong outing with her best friend, Kate. I was coming downstairs from the shower when the door opened. "Just dropping her off," said Kate's mother, and the two of us chatted for a moment while the girls sorted out their toys and said good-bye. Then Kate and her mom left and Rebecca came halfway up the stairs, to where I was sitting, to say hello. She put her arms around my waist and looked up at me.

"I heard you had ice cream," I said.

"Umm," she said, smiling. Those arresting green eyes, that lovely face. I smoothed her hair back. *What a knockout*, I was thinking. "It was soooo good," she said. "With *sprinkles*."

So this is where we have arrived—squarely at the crossroads of nature and nurture, enlisting all the tools science currently offers in the cause of beating back a fire in the brain that has smoldered in my family now for at least four generations. I can trace it back as far as that sad, taciturn woman who cut up her only winter coat so that her children would not be cold, who passed it along to the young mother who would sit in the red vinyl chair, silently weeping; who passed it along to me; and I in turn handed it down to that lovely child who, at the age of two, could say only, "I sad, too." It's a long road we've been on, and it's foolish to think it ends here. But there are times I let myself hope that, at least for my daughter, the road ahead won't be as hard.

It is by no means a foregone conclusion that children of depressed mothers will suffer from depression themselves; most, in fact, never will. But among the women I interviewed and who answered our survey, there was also clear evidence that many of their children were absorbing, as I did, ways of thinking inculcated by depression. When we asked, "Do any of your children ever say things that sound similar to what depressed people often say, such as blaming themselves for things that weren't really their fault, or being supercritical of themselves, not giving themselves credit for good things they've done, or thinking pessimistically about the future?"

42 percent said "occasionally" and 15.7 percent said "yes, often."

It was clear, for instance, that some of these children had already learned the depressive mantra: It's all my fault. (One mother told me that her stepdaughter's favorite T-shirt had those words on it.) At times, this even included feeling responsible for their mothers' depression. One mother wrote, "There were times when I was upset and angry at the world and [my son] would ask, 'What did I do?' " Another mother said she was sure that her children blamed themselves for her depression, even though they hadn't said so, "because I still blame myself for my mom's depression. You can't help but think if only I was a better or happier child, she wouldn't have worried so much [and] she would [have been] happier." Another mother realized only in retrospect that her child had been thinking this way. "When we went to the family therapist and I was initially diagnosed as depressed, the therapist made it a point to tell my son that he was not the problem," she said. "Immediately my son's combative behavior decreased."

Reflexive self-blame was also evident in more general ways. "My son, [who is] fifteen, blames himself if his team has a bad basketball game," one mother wrote. "He puts himself down and will go to his room and not talk to anyone. Then he usually cries if you get him to talk."

Then there's the assumption that any specific bad event is part of a larger, overall pattern, evidence of which also turned up. When her son was twelve, one mother told me, one of his friends mistreated him. "Instead of wondering what the friend's problem was, he wondered why people don't like him as much as he [liked] them." When he was fifteen, something similar happened, "and the same question arose: 'Why don't my friends like me as much as I like them?' " And there was this, from the mother of two teenage boys: "On Memorial Day we had a garage sale. The boys were told whatever they sold, the money was theirs to keep. [My younger son] was selling and made over ten dollars. [My older son] only made about a dollar fifty. [His] reaction was to feel the entire island of children hated him."

Then there were children who automatically assumed that bad things were permanent. One mother wrote that her daughter would frequently make remarks indicating that she thought their family would never get out of their financial bind—things like, "We're never going to take a big vacation on an airplane or a cruise, like other people" or "We're never going to fix up the house like other people's." Another mother described her children this way: "Even though they are all very bright, they claim they can't do well in school. They say they don't care what happens to them, they'll never amount to anything."

Another mother told this story: "My daughter is always sure she's not good enough, that she should be the best and never can be. She was asked to drop from the advanced chemistry course this past year because she failed the first two tests. She was sure that she was stupid and couldn't pass, but couldn't change her schedule without goofing up her graduation date, so [she] stuck with it, complaining, frequently crying. She passed the course with a B and was able to pass the test that grants college credit, something only a few students are able to do. Yet she minimizes this and talks about how poor she is in chemistry. She constantly fears a bad grade will cause her not to be accepted in college followed by a minimum-wage job and a struggle to survive the rest of her life."

A word that came up over and over in these women's descriptions of their children was *perfectionism*—coupled with a sense of failure despite high achievement. Here was vivid anecdotal evidence that a basic assumption that has driven the American educational system for the past forty years is flat wrong: achievement does *not* depend on high self-esteem. These were children who were achieving quite a lot, and who still believed they amounted to nothing.

"My oldest (eight years old) will sometimes have a fit and say 'Nobody likes me,' 'I wish I was dead,' or 'I can't do anything.' Yet she has many friends, is very bright, and is good at almost everything," one mother wrote. She would have a lot in common with another mother of an eight-year-old daughter, who described her daughter this way: "[She] believes she is fat. She is very hard on herself if she doesn't bring home all A's from school. If she messes up on the piano, she calls herself stupid."

"My daughter is a perfectionist. She always expects herself to be perfect and when things don't go the way she expects she gets down on herself quickly. She recently got her doctorate and [when she] did not get a position she interviewed for, she responded, 'Well, I guess I can go back to New York City and wait tables.'"

How could a young woman who has just earned a doctorate give up on herself after one defeat? The reason is the all-or-nothing thinking so characteristic of depression. A century ago, psychologist William James invented a formula: self-esteem equals success divided by expectations. But if your expectation is perfection, it doesn't matter how many successes you have, they will never exceed your expectations. The number on the bottom will always be bigger than the number on the top; you will always be a fraction of less than 1. That was how many of these children seemed to see themselves: despite their talents, intelligence, and attractive qualities, many of them showed little self-confidence, resilience, or belief in their own abilities. For these children, there was no such thing as partial success or a sense of value in having learned a skill; the only thing that counted was being the best. For boys, the yardstick was often sports; for girls, it was physical appearance. And these attitudes carried over into adulthood, except that for men career achievement or earning power seemed to replace the emphasis on sports, while women remained fixated on how physically attractive they were. The lives of these children were eerie echoes of the motherstress I described earlier: they seemed weighed down by impossible expectations, some of them self-imposed and some imprinted on them by a materialistic culture that worships physical beauty.

"My son has talked about not measuring up to his wife's expectations and he has gotten tearful about it," one mother told me. Said another mother, whose children were grown, "Our children set very high standards for themselves. One son who worked for a big auto manufacturer as an engineer was putting in twelve-hour days, many times [for] seven days a week, for ten years. His health began to fail. I think he blamed himself that he couldn't keep up with the demands of his company when really the company was asking him to do the job of three people. . . . One daughter is a writer in Hollywood. She would like to find a loving man but meets mainly men who are self-absorbed. Sometimes I think she thinks there is something wrong with her, yet she is smart, beautiful, kind, and caring." This mother might commiserate with another mother who wrote, "My daughter, who is extremely beautiful, will say things like, 'Why doesn't anyone like me?' or 'It's tough looking average.'"

Hearing these descriptions, I got the feeling that for these kids, an old aphorism like "It isn't whether you win or lose, it's how you play the game" would elicit snickers, or maybe just blank stares. One mother described her son's high school participation on the wrestling and swimming teams, even though he was not a star athlete. "He was always very critical of himself," she said. "He couldn't see that he has perseverance like nobody else." Memories of who won the county swim meet will fade; perseverance is a valuable asset for life. But what good will that asset do if he doesn't know he has it?

It probably seems unfair, maybe even overwhelming. If you are a mother who struggles with depression, how can you possibly be a decent coping-skills coach for your children? How do you teach children skills you may not even have?

You can't. On the other hand, doing nothing is not an option.

There are ways to help children who are in crisis, or who are at high risk for suffering from depression. But children and depression is a vast subject, largely beyond my expertise.

I do know this, however: no mother can help her children if she does not first help herself. No amount of therapy, books, or medication can be more powerful than the example set by a parent who comes to terms with her own illness and learns to manage it effectively. The best research shows that medication combined with therapy gets the best results in treating moderate to severe depression, and in recent years cognitive behavioral therapy in particular has been shown to be highly effective.

Still, as every behavioral psychologist knows, long-term change is rarely achieved overnight by dramatic pronouncements or some lightning-bolt epiphany. Change comes slowly, in small increments—each step cemented into habit through months of repetition and painstaking work. The road out of depression is like that. And before you start, you have to know your destination. It is not happiness. It is optimism.

If you are like me, that last word may trigger a knee-jerk distrust. Bookstores are full of feel-good solutions and ten-step programs to achieving all kinds of felicity—sexual, economic, health, parenting—and while many contain some useful advice, we all know that no one—not even Dr. Phil!—has a surefire fix for major life challenges. My first reaction when I saw the title of Seligman's best-known book, *Learned Optimism,* was to remember the old vaudeville routine where the patient flexes his elbow and says to the doctor, "Doc, it hurts when I do this." The doc says, "Then don't do that." *Oh, so it's simple!* I thought sarcastically.

I divided the world into pessimists, optimists, and realists, and prided myself on being a realist. (In some ways, I think I am; there's some interesting research indicating that depressed people do see the world more clearly than "healthy" people, being unburdened by any tendency to overestimate themselves or their ability to control events.) Optimists, in my opinion, were people who were too stupid (or maybe ideologically blinded) to grasp the nature of the problem. Eventually, however, my reading of Seligman's research (and, to some degree, my own experience in finding fulfillment in the work of writing) convinced me that what passes for optimism these days is not really optimism; it's Howdy Doody–ism. Authentic optimism is different: it's a combination of courage, tenacity, creativity, and resilience. The best illustration I can think of, believe it or not, comes from a war movie.

In *Saving Private Ryan,* there is a scene where a handful of American soldiers led by an Army Ranger captain (played by Tom Hanks) find themselves in a small French town. It is D-Day plus four, and they are far ahead of the main Allied forces, guarding a bridge that both the Allies and the Germans desperately need. It's likely that a German tank division will get there before the Allies do. The captain and his men are outgunned and outmanned. "Well," the captain says, thinking out loud, "we *could* use a sticky bomb." His men give him a look: a *what?* So he shows them: he crams some explosive compound into a standard-issue GI sock, shoves in a short fuse, and rolls the whole thing in axle grease. When German tanks arrive, the Americans wait until the tanks are almost on top of them, and then slap the sticky bombs on the wheels. The detonation knocks the tanks off their rollers, rendering them useless.

"Good old American ingenuity," I whispered to my husband in the movie theater. The scene stuck in my mind because at the time, Rebecca was two, and I was developing a newfound appreciation for ingenuity of the kind toddlers used to put themselves in peril, and the kind I was having to come up with to solve the crises that erupted every day. That stock phrase—"American ingenuity"—reminded me of another stock phrase, one that I often heard used almost interchangeably: "American optimism." Could it be that ingenuity—or creativity, or resilience, or, to put it another way, "thinking outside the box"—was an integral part of optimism? A pessimist in that situation would have said, "The odds are overwhelmingly against us; we should retreat." But the captain resisted that defeatist logic; he assessed the facts and worked with what he had. It was not a masterful gambit that routed the enemy, the way things work in action movies; it didn't even totally stop the enemy's advance. But it slowed them down, enabling the captain and his men to hold the bridge until reinforcements arrived.

The more I thought about it, the more it seemed to me that it was a good analogy to the effective ways I've found to fight depression. Depression never gets licked in a day; sometimes you just have to hang in there. But small victories are crucial. One small action here (deciding to get out of bed) combined with another action there (taking your medications, spending fifteen minutes with your child) creates synergy. Individually, no one thing will get you well—a fact that can be paralyzing. But put two or three things together, and the sum of the individual effects is mysteriously magnified. In a way, you could say an optimist is willing to think small.

This interpretation fits with Seligman's view, which deconstructs the optimistic explanatory style into three parts, just like the depressive explanatory style. When something bad happens, an optimist sees it as a temporary condition, not something that is bound to recur. When failure happens, an optimistic thinker attributes that to specific forces or events, not to some unalterable aspect of their personality, which means that things can change. They do not automatically blame themselves when bad things happen, nor do they instantly cast around for a scapegoat. When something is their fault, they don't assume blame in a way that amounts to character suicide; they simply own up to it, and then do what they can to make amends.

The tools to do this sort of thinking can be found in unexpected places. For me, it was journalism. When I suffered that deep, deep depression in 1990, I had been a newspaper reporter for thirteen years. Looking back, I can see that my craft had helped me to hone the kind of mental skills that helped me survive that hellish period. I loved journalism because every day was a different challenge; therefore, no problems were permanent. I learned to think of obstacles not as insurmountable barriers but as specific problems that were usually best dealt with by breaking them into their individual components. I'd done well in my profession because I had learned to assess my failures and learn from them instead of slinking off to find another line of work.

When I started applying some of these skills to my personal life—something I could do only when medication stabilized my mood swings—my life began to slowly but dramatically change. For me, a crucial turning point was when I began giving myself one task every day that was designed to help myself get better—whether it was exercise, or cleaning the house, or making myself go to a social event. If nothing else got done, that one thing still had to be accomplished. I relinquished any thoughts of magic cures; I resigned myself to setbacks. But I kept on doing my "one thing" every day, and over time, they added up.

Whatever your starting point, these are skills that can be learned. As one mother wrote, "You can teach yourself to unlearn old behavior patterns that adversely affect your situation. You just have to practice the new behavior until it feels *normal*."

"Every event in our lives has duality," said another mother in the survey. "Each can harm you or each can be learned from. . . . I've modeled for [my son] that taking your meds, not running out of meds, exercising, diet, meditation [and] prayer are all tools at my disposal for managing this disease. I've taught him that you shouldn't say, 'Oh I can't do X because I suffer from depression,' but say 'I can do X in spite of suffering from depression.' He's been given the tools. He can let them waste or he can build with them."

Finally, and not least, mothers with depression can make sure that their children have access to healthy role models. One theme that emerged in some of the survey responses was that sometimes one person can make a huge difference. Roughly 30 percent of the women who said they had grown up with depressed mothers said that someone—usually a relative, sometimes a teacher, friend, or counselor—reached out to help them. "My second-grade teacher encouraged me to write poetry and told me I was special and had a great future," one mother recalled. Another wrote that her grandmother "lived in the same city and would come to our house once or twice a week. She never really explained what was going on with our mom . . . but she always made it a point to get us involved in some activity, whether inside or outside the home. I think she even went to some PTA functions for us, too." Another said her biggest help during childhood was "my best friend, Sara. We have been friends now for seventeen years."

Sometimes all it takes is one person. "I no longer, for two decades now, refer to my stepfather as 'stepfather,'" one mother wrote. "He is my father, period. He was then and is now the greatest support I have ever known. Because of him, I have survived."

I am walking past Suzanne's room when I hear her inside, talking to herself. My four-year-old is wild about art, and lately she's been trying some ambitious multimedia projects. They usually involve about a pint of glue, half a ream of copier paper filched from my office, a dozen Popsicle sticks, a roll of tape, several different colors of paint, and about twenty minutes of cleanup on my part. But she is only four, so the effect almost always falls short of what she envisioned. At this moment she is frustrated and getting more so by the second. Finally she explodes. "I CAN'T DO IT!" she sobs. There is something about the way her face crumples that just unhinges me, and my knee-jerk reaction is to rush to her aid. But lately it's dawned on me that jumping in so fast is not helping her learn to handle frustration, and it's also undermining her self-confidence. She's begun running to me for help with things I know she can do—not a good sign. So now

I stand outside her door and wait. The next thing I hear is this: "NOBODY LIKES ME!"

Oh, boy, do I know this one. When I was in eighth grade and I threw my algebra book through my bedroom window (which was closed at the time), it was not because I had failed to understand algebra; it was because I *was* a loathsome failure. Now here's my daughter, an eerie little echo of the past. *How has she learned this?* I ask myself, and a helpful little voice in my head pipes up: *What about that time you backed the van into that lady's mailbox and swore at yourself and spent the next hour calling yourself 'stupid'? Suzanne was in the backseat when that happened. Or the time you dropped the egg on the kitchen floor, and called yourself stu—*

All right. I make a mental note to talk to David about this issue—but life is busy, I forget to bring it up, and before I get around to it, this happens:

Rebecca is in the family room tracing a picture of a horse head. Suzanne is trying to do the same thing, but her tracing paper won't stay put. David has gone to work and I'm getting breakfast for the kids and cleaning up the kitchen; soon, I'll have to get Rebecca out the door to school, and then Suzanne will go off to day care because it's a writing day for me. It's spring, and it happens to be a morning when I woke up with a pounding headache—allergies, probably. Suzanne's been fussier than usual lately and sounding very snuffly; I suspect she's suffering from the pollen as well. I notice that Suzanne persists for a while at the tracing, but gives up in frustration when hers doesn't look like Rebecca's.

The next thing I know, she's in her room, calling for help. Rebecca has just left for school and I'm heading upstairs to get dressed, but I take a detour to see what she wants. What I see is a piece of paper attached to the wall with several yards of Scotch tape. Glued to the paper are a couple of Popsicle sticks roughly in the shape of trees—very creative—and, next to them, a cutout picture of a horse. Suzanne is very agitated. "I want the tape off!" she says, pointing to the horse. Okay: I carefully peel the tape off the paper horse, taking care not to rip him in the process. When I show Suzanne what I've done, she falls to the floor, sobbing hysterically, "YOU TOOK ALL THE TAPE OFF!"

The sound of her screams feels like an ice pick in my skull. I am one nanosecond away from losing my temper—and then, I see two fat tears roll off her cheeks and fall on the floor. Something inside me uncoils, steps back. My headache, the need to get the day going, the mess she has just made—they all fade away, and in that moment I am able to see a frustrated little girl who probably doesn't feel well and who does not know how to say what she wants. I hold out my arms.

Oh, but it's not so simple. Suzanne is in a Force Five Swivet now—screaming, pounding the floor, red-faced, hysterically angry. I sit down on the floor and wait. I have learned to do that much.

Suzanne does not even want me to look at her, so I look at the floor. After a few more minutes of incoherent rage, I see her out of the corner of my eye going over to her bed, where she grabs a teddy bear. Then she crawls under her drawing table with her bear. Her sobs are starting to die down. After a moment, she gets up and takes her teddy bear over to the rocking chair, where she curls up. Her sobs are now more like hiccups; she is calming down. Finally I hear a muffled voice,

"I want my boppy"—her pacifier, which she still depends on in moments of crisis.

"Okay," I say. "I can go look for a pacifier, if you want me to. But that means I'll have to go into the kitchen." No reply. I stand up—and that's when she is finally ready to come to me, arms outstretched: pick me up.

I hold her close for a few minutes. Suzanne likes to put her cheek against mine, and I love the feeling of that silken, perfect skin. At this moment, though, that perfect skin is slippery with tears and snot. She wipes her face against mine, smearing muck all over my cheek. I don't care. There is something we both need about that touch, some communion that takes place at a level deeper than words. We stand like that for about two minutes. Then I start talking softly in her ear.

"It is so frustrating to be four, and to want to do things you're not quite able to do yet but you see your big sister doing, isn't?" Short nod. "And sometimes you can't always say what you really want to say, and sometimes on top of that there are days when you just feel rotten." I pause. She makes no move or reply; she is listening intently. "But you are getting bigger and better at things every day. This time last year, you could never have cut out that horse, or glued those Popsicle sticks to look like a tree, or colored around them the way you did. Every day you are learning more and more and getting better at doing things you already know. It won't always be like this, sweet pea. You are learning so fast, and I love you so much. And you know something? I really liked the fact that even though you got so upset, you figured out a way to calm yourself down. That is just so cool."

The effect of my words is magical. She wriggles out of my

arms and stands on the bed, so that she is able, for once, to be taller than me. "But, Mommy!" she says exuberantly. "I'm better at frog noises than Daddy is!"

Someday, my children will leave me, too. It's a dreadful prospect. Unlike my mother, though, I realize that there is something worse: raising children who think they *can't* leave, who lack the self-confidence to test their wings. It has recently occurred to me that Rebecca is eight, already halfway to her driver's license, and yet it seems we put the baby crib in storage just the other day. We are hurtling through childhood at warp speed, and there is so much my daughters need to know.

But one of the many great things about children is that they can learn from your weaknesses as well as from your strengths. I cannot give my daughters boundless self-confidence, because I've never known what that feels like. But I can teach them what I've learned about overcoming habitual self-doubt, and about hammering out a belief in themselves based on actual accomplishments. I have a long list of mistakes they can learn from. And when I fall back into my "default" position—the old depressive ways of thinking that I learned so early in life—I can show them that it is possible, though painful, to crawl out of that hole.

And then, when my children run into some problem that seems insurmountable to them, I can share some of my tools with them. Together, we figure it out.

CHAPTER SEVEN

Coping

Some days the TV and videos help keep the kids occupied. . . . I don't think it is healthy for them to watch too much TV, but some days it is healthier than them interacting too much with me. (Mr. Rogers is a lot nicer than I am when depressed.) The most valuable thing has been finding other people to talk to, especially other mothers. The Internet helped early on because it didn't make me face people. . . . Parenting bulletin boards were a great source of encouragement.

—ANONYMOUS SURVEY RESPONDENT

Emily is a Presbyterian minister who presides over a medium-sized congregation in a suburb south of Baltimore. A slender woman of thirty-seven with sandy blond hair and a runner's angular physique, she is married and the mother of a five-year-old son and a three-year-old daughter. When I first interviewed her, in the summer of 2004, it was to talk about a severe episode of depression she had suffered two years earlier, when she had made the difficult decision to take antidepressants. At that point, her depression was what psychiatrists call "in remission"—but, given her history, her doctor cautioned her that she would probably have to be on medication for life. (The episode in 2002 was her fourth.) It was clear that she would much rather be living life without pills. But just as I was getting ready to leave, she mentioned, almost in passing, that she was a little worried. Lately, she'd been noticing a few little things—nothing major, she said, but enough to make her wonder if she was headed for another episode. The remark stuck in my mind.

I find something to like in most people I interview, but there was a special connection I felt with Emily. I could tell she felt it, too. We had some important things in common— the ages of our children were similar, we lived near each other, we had similar interests in books. With those things, and that passing remark at the end of our first interview, Emily stayed on my mind. I wondered how she was doing. So in the spring of 2005 I manufactured a reason to call her again.

Our second interview took place just before Easter 2005. The air was still chilly, but spring was just offstage, awaiting its cue: outside, the red maple buds were swollen, the cro- cuses had emerged. Emily was wearing brown slacks and a purple turtleneck, with a large amethyst crystal on a gold chain around her neck. She looked stylish and composed. But as it quickly became clear, looks were deceiving.

"It's been a hard winter," she said. She might have been speaking about the weather, except that we both knew she wasn't. Then I noticed her hands. She seemed unaware of them, but they were constantly in motion, twisting nervously around each other in her lap. Here was a woman who was, literally, wringing her hands.

The decline had been slow. From one day to the next, the misery index ramped up in such small increments that she just adapted—but by the fall, she had given up her running schedule, even though exercise had always been one of her chief pleasures. She had no energy for it. Instead, she had taken to cocooning on the sofa at night, snacking on sugary foods, crackers, chocolate. This craving for carbohydrates is a common form of self-medicating for people with depression, since eating carbohydrates stimulates the production of serotonin. It can be a vicious cycle: the increase in serotonin levels is temporary, but for most people the calories are forever (and the extra pounds become another reason to feel depressed). Emily was lucky: her metabolism burned up everything she ate. Still, it was another one of those troubling symptoms.

Mentally, she seemed distant, just "not there" to her husband and children, five-year-old Sam and three-year-old Sara. If someone had asked them, which later her husband did, they would have said she was "crabby." She wasn't sad. What she felt was the slow draining away of ordinary pleasures, which started as a kind of numbness and then, as the weeks passed, progressed toward a kind of cold mental turmoil—anxiety, combined with a constant interior drone of caustic self-criticism ("the demons," she had called them in our first interview, describing the intense self-loathing depression inflicted on her). This time, she kept telling herself that she was just having a couple of bad weeks. She was trying to ignore something that would not ignore her.

And yet for short periods, she could operate with half a brain. She could, for instance, mentally split off from the part of her that was in pain long enough to preach a sermon, or to teach her scheduled class in Christianity. "For those forty-five minutes," she said, "the demons were quiet." This ability to divorce herself from her pain was useful in the short term, but it also created barriers. Seeing her speaking confidently in the classroom or in the pulpit, no casual observer could have imagined her mental anguish. Only three people knew: her husband, a close friend in Ohio who could read every nuance of her voice over the telephone, and her running buddy.

Reluctantly, she went back to her psychiatrist.

Throughout that bleak winter, as she waited for the change in medications to have some effect, she found refuge in music and poetry—specifically, the Book of Psalms. It's the hymnbook of the Old Testament, a collection of poems and songs written several centuries before Christ by different authors, some anonymous. A surprising number are starkly modern portrayals of emotional desolation. "Be not far from me; for trouble is near; for there is none to help," says one. "Deep calleth unto deep at the noise of thy waterspouts; all thy waves and thy billows are gone over me"—an eloquent description of the crushing power of despair. Or this: "From the end of the earth will I cry unto thee, when my heart is overwhelmed / Lead me to the rock that is higher than I"—an abject plea for protection attributed to none other than David, powerful king of Israel. In this ancient poetry, she found the echoes of human voices from thousands of years ago, suffering then as she was suffering now. It helped, she said, "to know that I was not the only one, that somebody before me had felt this way, too."

The days passed; winter faded. And slowly she began to find small bits of pleasure in her day, enough to convince her that pleasure was still there, that it had been there all along. Her faith had not given her any miracle cure. All it gave her was a sense of companionship in the darkness. But it was enough.

It's hard to say which is more difficult: holding down a job outside the home while suffering from an episode of serious depression, as Emily did, or being a stay-at-home mother fighting the struggle. But whether or not they are members of the salaried workforce, all mothers with depression face the assumption that they will continue to care for their children, no matter what. *Mother* may be the only word in the English language where its use as a noun and its use as a verb are so closely linked. Once you *are* a mother, you mother—or are on standby to do so—twenty-four hours a day. How do mothers with depression cope?

Some of the answers I heard from mothers were the classic self-destructive ones: drinking too much, abusing prescription drugs. Many women described just toughing it out. Others seemed to have an instinctive understanding of some key elements of cognitive therapy. "Serving others is quite helpful in battling depression," one mother wrote—a response that, had she known it, is backed up by research that shows that identifying a particular skill—empathy, in her case—and using it well is a source of deep emotional satisfaction. (It also

builds on an old precept taught in Alcoholics Anonymous—
"fake it until you make it"—which, in turn, resembles a key
concept of cognitive therapy: actions can create emotions just
as emotions create action.)

Some women found ways of coping that involved spending
more time with the children, but most described ways that involved withdrawing from family life. In the culture of Intensive Mothering, it's tempting to automatically label the first as
"good" and the second as "bad," but things aren't so simple.
Very often it's not *what* mothers did, but why and how they
did it, that made the difference between effective coping and
simply escaping from life.

The coping skills these women described to me are not tri-
umphal epiphanies. They're a bit like what you might hear
from someone who has survived, say, a shark attack: they are
practical and concrete, and they bear the imprint of authentic
experience. No one coping technique is likely to bail you out,
and some are examples of what *not* to do. But somewhere in
here, you are likely to find something useful.

Prayer, Meditation, Reading, Keeping a Journal

"Withdrawal" can signify a mother's emotional isolation from
her child, or it can simply be a creative response to stress. "A
room of one's own," in Virginia Woolf's phrase, is not just a
metaphor for economic or artistic independence; in times of
emotional turmoil, for some women, solitude is as much a
need as food and water.

"I turned my bedroom into a retreat [with] candles, soft
linen, meditation, etc., with the door closed [and] locked,"

one mother wrote. Many other women mentioned such solitary activities as reading, listening to music, or keeping a journal. The difference between solitude and withdrawal lies in whether the time is used to escape reality or to confront it. "I tend to spend more time in my bedroom surfing on the Internet or playing computer games," wrote one mother; for her, withdrawal was just a chance for a mindless escape. (Not that this is all bad; once in a while, we all need to play computer solitaire.) But for another mother, time alone in the bedroom was a chance to "[keep] reading and learning as much as possible as quick as I could, to progress forward into change. I knew it would get easier, so I [tried] to find out when that would be and make it happen as quick as I could."

Other mothers mentioned prayer. "I pray a lot," one mother wrote. "Some days it is from the moment I wake to the moment I go to bed, one long conversation with God." Other types of meditation are also effective. "I have recently started taking yoga once a week as a way to decrease my response to my depression, and I try to use breathing exercises when my anxiety increases (not always successful, but it is at least a starting point)," wrote one mother. Added another: "Since I have been in recovery and on antidepressants (three years now), my depression is under control. If I have times when I feel a little blue or overwhelmed I ask for some quiet time from my husband. It has taken a long time but he now understands how important this is for me."

People who do not approach depression as a spiritual issue may not find these responses helpful, and even for those who do, the answers made it clear that prayer is not the magic cure some well-meaning religious people say it is. "In the beginning of the most recent episode, I also prayed and meditated a lot, but nothing seemed to help. I felt God had deserted me," one mother wrote. Depression can force an examination of beliefs that have been in place since childhood, and those beliefs don't always hold up to scrutiny. Emily, for instance, acknowledges that for her, depressive episodes are spiritual crises, but she is irritated by the insistence of many mainstream pastors that depression is *only* a spiritual issue. "People forget," she said. "We are spirits in *bodies*."

Barbara Hinson (not her real name), a thirty-two-year-old mother of a boy and a girl, is a parishioner of a predominately black Baptist church in a community south of Baltimore, Maryland. When she began suffering from severe depression and panic attacks in the summer of 1998, a few months after the birth of her second child, her pastor told her that all she needed to do was pray. At the same time, her doctor was urging her to go on antidepressants.

Such conflicting advice "makes you question your faith," she said. "You think, 'If I take the medicine, does it mean I don't have faith God will pull me through this?' " It's a question faced by many women raised in religious households where the emphasis was put almost exclusively on the healing power of prayer. Black women often face the added cultural stereotype of the Indomitable Black Matriarch. As one mother from North Carolina told me, "We're supposed to be superwomen—we handle everything with the greatest of ease."

Pushed one way by her pastor and upbringing, pulled another by her doctor, Barbara vacillated for months. Then, one day in her pediatrician's waiting room, she struck up a conversation with another mother who casually mentioned she was taking medication for depression. Barbara went home, started to cry, and couldn't stop. In retrospect, she realizes that it was only by connecting with someone who had a similar problem that she was able to begin to face the reality of her own—a reality that was at that moment overwhelming. Until then, she had simply been a mother with two children, and "I didn't see having two children as a problem." Like me, she had persuaded herself she wasn't ill, merely incompetent. It still took her several months to overcome her fear of medication and to find a drug that could help her (Zoloft made her slightly manic, but Paxil worked).

Her story has a sequel. Not long after this, she was asked at a church service to "give a testimony," which, in her religious tradition, usually means a short speech praising God for deliverance from sickness or trouble. Instead of giving the standard testimonial, Barbara decided on blunt honesty. "I was unsure of the reaction," she said. "I thought, 'Maybe I'm an outcast already.' " But after she told her fellow church members about her struggle and her decision to go on medication, she was amazed at the response: several people sought her out after the service to tell her they had been through the same thing. Later, her pastor told her he had found her experience educational, and he now realized that his advice on the subject of depression had been off the mark.

"I Just Head for the Bed"

Obviously, sleep can be a form of escape. "When I am having depressive episodes, I long to just sleep, and I do," one mother wrote. "It's become a kind of joke in my immediate family. When I lay down my husband and kids joke that I'm 'going north.' " How do you know if you are using sleep as an escape? One danger sign is if you are consistently using drugs to induce it. Some of the women I heard from said forth-rightly that they used drugs or alcohol as a way of getting to sleep, but there were many more who hinted broadly at the same information. "I have slept forty-eight hours straight to avoid interaction and/or activity within the household," one mother wrote, for example, and it is hard to imagine doing that without some kind of pharmaceutical assistance. Over-the-counter drugs work just fine, as this response illustrates: "I tell [my children] Mommy has a headache, so I'm going to take a loooong shower, and a little nap (three hours) after tak-ing Tylenol PM."

On the other hand, exhaustion is a classic symptom of de-pression, and when your body is depleted and exhausted, sleeping is the only sane response. Even in times of relative well-being, people who battle chronic depression find that getting enough sleep is necessary to keep depression at bay. Emily, for instance, describes her own need for sleep as "non-negotiable" in good times as well as bad. There are ways to minimize the impact on children of their mother's extra need for sleep. "I would often take my daughter to school and then go back to bed," one mother wrote. Others seized the oppor-tunity to nap while their husbands were there to oversee the kids, and others contrived ways to nap while their children were napping. One mother even found a way to enlist her children's help: "I would set the timer on the microwave for twenty minutes and nap on the couch," she wrote. "My girls would watch a video or play on the floor next to me. I told them their job was to wake me up when they heard the timer go off. They did great! I got in a huge number of naps this way and my girls felt involved."

Keeping Busy

The best ways of staying busy involve not just getting some task accomplished, but using that task as psychological leverage to lift the spirits. "I was lucky enough to have a mother who understood [depression] and would not let me just lie around all day and cry," one mother wrote. "She would come to my house with a plan of what we were going to clean and we would do it. At the end of the day, I felt better to see what I had accomplished." Whether she knew it or not, that mother was using two highly effective cognitive therapy techniques: breaking down large tasks into smaller tasks, and then writing things down so her daughter would have a reality check about what her day had really been like. (A depressed person is capable of putting in a hard day's work while still maintaining "I didn't do anything today" because, at that moment, whatever she did felt futile and devoid of meaning.)

Another mother wrote that she was able to motivate herself by setting small goals at least some of the time: "I make a list of goals that I want to get done [and] I set one for each day," she wrote, and then added, "But when the depression hits hard, I sit at home and eat." Another wrote, "I make myself busy. I clean the playroom, do loads and loads of laundry, pay bills." If nothing else, these mothers were avoiding passivity, relieving some of the stress of the interminable to-do list, and, possibly, creating time for fun later.

For me, being physically active has always been a relief valve, and over the years I've come to recognize that an urge to clean up, clear out, and throw away is often a prelude to a period of very productive work. I tell my friends that when I'm not feeling great, I rearrange the furniture; when I'm really depressed, I head for Home Depot. Once, when I was still working at the *Washington Post* and was too depressed to read or concentrate, I called a friend who was about to have her bathroom renovated. "Let me strip your wallpaper for you," I said. "Are you sure?" she asked; she was perfectly willing to pay a contractor. I was sure. She saved money and I got some manual therapy. It helped. Every time I look at the walls of my study, which are painted a deep, relaxing shade of taupe, I think of that hot July several years ago when paintbrush therapy helped me stanch the despair in my mind.

But, like everything else, housework can morph into pathology. Over the years I've caught myself with a toothbrush in hand, scrubbing the grout between the bathroom tiles, on my hands and knees pulling up tiny pieces of clover on the lawn, or, on one memorable occasion, preparing to scrub my laminate kitchen floor with bleach, while in the next room my husband and the kids were having a fine old time romping around. At such moments, a tiny sane part of my brain has interrupted to ask, *What the heck are you doing?* and I've been able to stop. When housework becomes so compulsive that even *I* recognize it's a problem, I know I have been handed a warning sign. A belated recognition is better than no recognition at all: one mother described how she "worked long and hard, trying to make anything and everything better. I felt such chaos in my mind that I kept everything on the outside clean and orderly, neat and in order." She kept doing this, she wrote, right up to her suicide attempt.

So staying busy can be good, or it can amount to just aimless activity, or, as one mother said, "[filling] our days with endless activity so there is no downtime for my sadness to show." Another wrote, "I drag [the children] all over the place so that they are kept busy and not driving me crazy or getting on my nerves."

Shopping was a frequently mentioned form of pointless busyness: "I like to go to town every day in the morning—for crap, really," one mother wrote. "I can spend close to forty dollars on that." Wrote another: "I go shopping, which of course puts me into debt, which makes me more depressed." Another wrote: "I took [my children] to Target or Wal-Mart and we would all buy something new. I tried to make up for the guilt that I felt for being depressed."

But it's possible to recognize that pattern and change it. "I have always used staying busy as a coping mechanism, often leaving the house and going anywhere to avoid the way I [felt]," one mother told me. "Now that I understand my depression a little better, I will sometimes use the bad days to just chill out and snuggle with my kids. We have a very low-key day playing or watching their favorite programs on TV . . . I allow myself to take a 'mental health' day off."

Having a daily plan helped some mothers; others found it easier to just go with the flow. "I've tried to vary activities and locations. For example, instead of staying in the house all day, I'd have my child help me with one chore, then we'd go outside and let her run through the sprinkler, then I'd ask her to watch one TV show while I took a much-needed nap, then we'd take a ride to get a little snack or something, then we'd play cards or do an art project," wrote one mother. "Just keep the ball moving."

The Electronic Sedative

The work of motherhood, as I've noted, involves the willingness to step away from self-absorption and toward a vigilant attentiveness to the needs of an "other"—and a central feature of maternal depression is that this shift in focus often becomes acutely painful. For many mothers, that's where television comes in. I expected to read a lot of comments about how mothers used television as a babysitter, and in fact thirty-seven mothers noted specifically that this was their usual way of coping ("I allow my child way more TV time than normal," was one typical comment). What I did not expect was that in roughly half of the responses involving television, the person getting more tube time was Mom (though possibly this meant more TV time for her children, too, in another room). These mothers described television as a kind of electronic cocoon that muffles the noise of the outside world for them, a way to at least temporarily divert their minds from the well-worn circular path of their thoughts.

"If I could concentrate on a TV show, I would try to 'lose myself' for a little while," one mother wrote. Another wrote, "I have spent many a lazy day in front of the television just trying not to think. Just zoning out. Eating eating eating, cleaning, sleeping, watching TV—these are things that are somehow sedatives to feelings. I just shut down while doing these things." Responses like this raise some interesting questions.

Are the children in these households being exposed to adult programs they might not otherwise see? There has been ample research linking childhood obesity with excessive television viewing; do mothers who spend too many hours in front of the television also tend to eat compulsively? (It certainly seems possible.) Are these mothers using television to escape their children, their husbands, or both?

Even here, though, it was possible to find examples of how to use television in a way that was at least not destructive, and may actually be helpful: two mothers said that during bad times, they planned TV watching with their children, "so that we can all escape reality for a while," as one put it. I've done this myself, and because I am a fan of the art of animation, cartoons are my visual medium of choice. My older daughter and I love the deadpan humor and urban sensibility of *Hey, Arnold!* on Nickelodeon. But mostly I haul out the old stuff: our household has a four-disc CD set of Looney Tunes cartoons that I put on when I can't muster up the energy to be with my children any other way. They're animation classics, they're fun, *and* they're educational: where else are today's children going to encounter references to Richard Wagner, Salvador Dalí, or Gioacchino Rossini? The Blake Edwards *Pink Panther* movies, with the immortal Peter Sellers, hold up well in the hilarity department, and my kids adore the sight gags. *America's Funniest Home Videos* is uneven, but if you're willing to sit through a whole hour, there's usually at least one belly laugh. Sometimes we turn the sound down and invent our own commentary, a trick I learned from my father, who used to watch old B-grade World War II movies this way. ("You know, if you stick your head up out of that foxhole, you're gonna—okay, now see what happened?") Even mediocre movies can be funny if you watch them with the sound off. ("Who's that guy, Mommy?" "The bad guy. They're the ones with the slicked-back hair who never look behind doors.") The point here is to find something that delivers relief for you and entertainment for your children. Laughter really is healing. I used to know a D.C. paramedic who would recover from the blood and gore of his nightly shift by watching old episodes of *Darkwing Duck* and a child psychiatrist who swore by the restorative powers of *SpongeBob SquarePants*.

So the boob tube can be useful. But I also understand why two mothers in the survey said they turned the television off to help them cope: when I am struggling with depression, the usual news diet of war, crime, and natural disaster puts me deeper in the hole. A week after Hurricane Katrina hit, I had to increase my anxiety medication; when the Washington, D.C., suburbs were terrorized by random sniper shootings in the fall of 2003, and one of the shootings happened roughly a mile from my house, my stress level skyrocketed and became a major depressive episode. If you are supersensitive to stress, even secondhand exposure to tragedy can be unhealthy. It may sound callous to turn *off* the news, but look at it this way: to help anybody else, it helps to be healthy.

Getting Help from Relatives and Friends

Simply talking to others (sisters and mothers seemed to be favorite confidantes) was a good, everyday kind of relief valve for many of the women I talked to. Some women found this outlet in joining a playgroup, or just getting together with a friend. "The most useful thing for me was to get together with my sister-in-law and her kids," one mother wrote. "That way, the kids would play and we moms could sit down, enjoy our coffee, and share experiences and advice."

That sentence points out a crucial distinction, which is that the talking in question does not focus on the depressed person's feelings. It's an article of faith in our culture that ventilating our feelings is a healthy thing; for many years, the whole psychoanalytic industry was built on this premise (not to mention countless soap opera plots). But when it comes to depression, an illness characterized by brooding and circular reasoning, a little talking about your illness goes a long way. The poet John Berryman, who spent a fair share of time in mental institutions, once wrote a piece of doggerel: "Staring at the ceiling will / in time, induce a mental ill." He was right. Endless discussions about how bad you feel will eventually make you feel worse, and besides, after a while you may discover that your confidante may have suddenly discovered a series of pressing appointments on the other side of town. When I am feeling really rotten, I find myself trying to navigate a course between letting people close to me know, fighting the urge to withdraw completely, and fending off the desire to indulge in a rant about my stinking mess of a life. It's a difficult line to walk. Sometimes you do need to reach out, but it helps to make a short list of people you can depend on who understand what you need. One of the best acts of friendship I've ever experienced happened some years ago when Rebecca was only nine months old and I unexpectedly got pregnant. In the space of seventy-two hours, I went from horror at being pregnant again so soon, to developing signs of an impending miscarriage, to a desperate desire to keep that baby. When things began to look grim, I finally called my friend Jean. "I'll meet you at Borders," she said. We did not discuss my problem; we just spent an hour or so looking at books with our babies in their respective strollers, and when I went home that day, I was able to deal with the bad news. *That's* a good friend.

The best confidante may or may not be your husband; in fact, the women I talked to were pretty evenly divided on this issue. Spouses of depressed people are themselves at a higher risk of depression, and husbands or partners may themselves need someone to talk to; there are times when expecting them to be your support is simply expecting too much. I have been forced to recognize that David is not going to be able to offer me unconditional support all the time, and I actually think that's healthy: after all, the first priority is to keep the family ship afloat, and *somebody* has to man the bridge. But I've also found it hard to explain to him exactly what feels so awful. Like many people who have never experienced severe depression, there is something about the experience that remains inaccessible to him. Other women found this to be true of their spouses or partners, too. "I talk with my husband about it often although he doesn't really understand what I'm going through," said one. I finally came up with a way to get it across to David one day when I said, "Try to imagine going to work and having to do your job the same as always, but with 30 percent less IQ." He had been hung up on the part about being sad for no reason; when I put it in cognitive terms, it made more sense to him.

There will be times when you will need support from out-side your family, and a friend can be a lot more helpful than a close relative who is suffering collateral damage. "My family is no help; they also (besides my seventeen-year-old son) think that depression is a joke [and] not real," another mother wrote. "My friends are the ones who get me through." On the other hand, understanding support from your life partner is not something you can duplicate anywhere else. As one mother wrote, "My husband, who *finally* understands, is very helpful. I rarely want to leave the house. He gets me out."

As with other coping methods, there was a way in which asking for help from friends or relatives could be described as unhealthy: a few mothers said that when they were de-pressed, they sent their children away to relatives, either on a regular basis or for long periods. "Every weekend I would send my children to my parents' house," wrote one mother. "This went on until they became teenagers." Making a habit out of asking for this help seemed to indicate that everybody, includ-ing the mother, assumed the situation would never get better (although obviously the care of a kind relative or friend may well be better for children than constant exposure to a se-verely depressed mother).

Abusing Alcohol and/or Drugs

Decades of antidrug rhetoric notwithstanding, it is a fact that illegal drugs, painkillers, and alcohol *work*. Entire Third World economies and vast First World fortunes are built on the fact that these substances offer a chemical womb to crawl into where, for a while, everything is rosy. Their effects are temporary and their hazards can be life-threatening, but the promise of surcease from some of the most overwhelming emotions of depression presents a powerful allure. Many mothers spoke of the ongoing struggle to yield to the craving to drink or misuse drugs. "I have tried many drugs," one mother told me. "I guess that is one of the hazards [of] being depressed [is that] you grab what you can."

And lots of women do. Nearly half (41 percent) of the mothers who responded to the survey said they had at some point dealt with their depression by using alcohol, misusing prescription drugs, or using illegal drugs. Aside from alcohol, the list of illicit, prescription, and over-the-counter drugs that came in for mention ran the gamut: cocaine, Vicodin, codeine, Tylenol PM, Xanax, Klonopin, Benadryl, over-the-counter diet pills, speed, hashish, heroin, and marijuana. A follow-up question about how helpful that had been elicited some gallows humor: one mother said that, for her, abusing alcohol works "pretty good—if I drink enough, I pass out." Another responded cheerily, "Pot helps a lot!"

But there was a general consensus among the mothers I talked to: they agreed that drug or alcohol abuse was not a viable long-term option. One mother said that marijuana "exacerbated my depressed feelings and seemed to open a floodgate of negative thoughts." Another mother said that drugs and alcohol "helped for an immediate fix, but my problems multiplied as a result. I definitely had to stop using/abusing them first before any legitimately prescribed medication would work for me." The long-term problems spawned by the misuse of drugs or alcohol includes worsening depression, physical withdrawal symptoms, increased anxiety, memory impairment, addiction, and, not least, guilt. "They just make the problem SSSSSOOOOO much worse" was one typical response. "You feel better for about fifteen minutes, and then it is guilt city."

Anxiety, more than depression, has given me a visceral understanding of the roots of drug abuse. For me, anxiety is the psychic equivalent of being swarmed by red ants, and it creates an almost overwhelming temptation to do something—anything—to relieve that feeling. The problem is that a lot of antianxiety drugs, unlike antidepressants, work almost immediately. That creates what addiction specialists call the "feedback loop," in which the brain quickly learns to expect that relief. That spawns a psychological dependence, which then frequently leads to physical dependence. Xanax, for instance, delivers within ten minutes—a gentle, limber warmth not too different from the effects of a good stiff scotch. I like it too much. Over the years, I've become seriously dependent on it several times, and had to suffer through the nausea and splitting headaches of withdrawal. I've done other bad things. I have hoarded pain medication prescribed for outpatient surgery or dental procedures so that when anxiety hit I could use it to get numbed out. There have been times when, if nothing else was available, I've just swigged down a staggering dose of NyQuil. For years, I abused alcohol to deal with anxiety. I would probably still be doing that if it weren't for the fact that my body's ability to deal with alcohol has become almost nonexistent, and now it just makes me sick. I miss the melting away of tension, the freedom I used to feel to go that third round in convivial company, the ease of expression it gave me. Still, I think my body's reaction to alcohol is a pretty clear message from the cosmos: *Your chemistry experiments are going to kill you.*

Exercise

There are any number of clinical studies that show that exercise is a valuable way of alleviating the symptoms of depression. They show up in the newspaper from time to time, always reported with an kind of cheery "good news!" twist. I suspect that a lot of mothers who suffer with depression find this to be . . . well, depressing. For them, exercise merely represents one more area of life in which they see themselves as failures. Mothers find it hard to find any time for themselves, and depressed mothers have a hard time finding the time *or* the motivation.

My own exercise habits are best described as "consistently inconsistent." When I was growing up in the South, only idiots, farmhands, or members of the varsity squad actually *volunteered* to run around in the oppressive heat that defined roughly nine months of our year. For the rest of us, the goal was to *avoid* sweating. It's an attitude that still holds sway in much of that part of the country (Southern states have some of the highest obesity rates in the nation) and I don't suppose I'll ever be able to completely rid myself of my early exposure to the concept that sweat is bad. I still don't particularly like exercise. I do, however, like the feeling that exercise gives me, and I miss it when I don't do it. And I am motivated to keep trying by the example set by my mother, who never learned the value of regular exercise and whose health has suffered greatly as a result.

```
      #300  06-05-2012 6:39PM
  Item(s) checked out to 22140001168630.

TITLE: The ghost in the house : real mot
BARCODE: 33504004935112zg031
DUE DATE: 07-05-12

TITLE: Which side are you on? [sound rec
BARCODE: 32140003429210
DUE DATE: 06-26-12


         What do you GEEK?
   www.howelllibrary.org/GEEK.htm
```

Here again, it helps to think small. There was one person in my youth who introduced me to the idea of exercise as pleasure, and that was my father, a natural athlete who tried to teach me tennis and golf. In my hands a golf club was a lethal weapon, but tennis was fun. I spent a lot of time batting balls against the brick wall in back of the Kroger grocery store when I was growing up (there must be a couple hundred rotting up there right now). Today, despite two bad knees, I can still hit a decent forehand. So: some exercise is better than no exercise. You don't have to start training for the Marine Corps Marathon; start by walking around the block.

Finding an exercise buddy can be extremely useful. I lost ten pounds once without even trying when a neighbor teamed up with me to go for walks early every weekday morning. There were times when Cindy would stand at the foot of the stairs and shout, "Get out of bed!" and there were times when I would do the same for her. Individually, we would have never stuck with it; together, we walked off the pounds. Variety helps, too. In the summer, I swim; in the fall, I walk; in bad weather, we have an elliptical exercise machine. (I would rather spend half an hour in the dentist's chair, but it's better than doing nothing.)

One mother I heard from found lifting weights to be therapeutic—and, since she lived in the country, so was taking the family dog on regular walks. "I allow myself to walk to the end of our road with a heavy heart and a head full of worries. However, when I head back home, I concentrate on the beauty around me—the hills, the fields of yellow and blue wildflowers. I breathe slowly in and out to relax, and listen to the calls of the Whiskey Jacks and blue jays." Almost any kind of physical activity helps when the brain is not working, and it doesn't have to be aerobic. Baking bread did it for one mother in our survey; another mentioned doing yard work.

Why does exercise help? One theory is that sustained physical activity ramps up the body's supply of endorphins, inducing the so-called runner's high. (I like getting high as much as anybody, but my body evidently requires a Clydesdale-sized endorphin dose to achieve an exercise high; it's never happened for me.) I do know, however, that a brisk walk almost always helps me think better. There is something useful about a simple change of scene. "It may sound strange, but we would often go outside," said one mother. "I am usually not self-aware enough to figure out that this will help, but we would end up out there by accident (to check the mail or get something out of the car), and I would realize that the sun and fresh air made me feel better."

Escaping to the Kids

Most depressed mothers I talked to said they needed more of a break from their children when they were depressed, but some felt the opposite. For those mothers who derived solace from being around their kids, it was clear that the interaction helped everybody. One mother, for instance, told me that what helped her most was "throwing myself into something—whether it [is] cleaning a room or playing ball with the kids. I always try to include the kids as best I can. They lift my spirits and lighten the load." Another described "playing with [my children] and watching the sheer delight of life that children have. Doing things like going to a park and swinging and playing in the sand helps a lot." Another mother who found it too hard to be physically active wrote that she found it helpful to rent funny movies and watch them with her kids. Still another wrote, "Having my children around helps when I'm feeling depressed. I love them so much and they are so darn cute, sweet, and funny." The theme in these responses is the abandonment (if only temporary) of the self, a crucial first step on the road to getting better.

I would love to be able to report that this is an unfailing spirit-lifter for me, but it's not always true. I am by nature a driven person who finds it hard to relax, and when I am depressed I am even more so. I am not the mother you see pausing with her little girl to smell the neighbor's roses; I'm the one standing in the doorway in her pajamas at 8 A.M., screaming at her child to get her butt out the door. Depression takes away my margins. I envy mothers who find relief in playing Chutes and Ladders, but I find it hard to be one of them.

On the other hand, having children who are still young enough to cuddle has taught me the immense healing power of touch, the calming effect that the warmth of your child's body against your own can have. Sometimes, if my children will let me, I will just snuggle on the sofa with them for a half hour or so, watching the dreck of their choice while I get a small endorphin lift. (This also helps me monitor what they're watching.)

A few responses described mother-child interactions, however, that seemed to involve an unhealthy blurring of boundaries. One mother wrote, "[I hold] her closer during the moments I want to flee the most. I rock her back and forth, praying out loud, thanking God for her life." Another wrote that "playing with the children . . . helps me stay focused on having a reason for being." These responses seemed to me less about the mother's ability to lose herself in child's play and more about children being cast as secondary characters in the mother's ongoing drama. Being another person's reason to live can be a heavy burden. And what happens if, while you are rocking and holding your child close, he would really rather be roller-skating?

Counseling and Medication

The medical establishment's party line on treating depression is that both counseling and medication are the best way of treating moderate to severe depression. Yet only fifteen of the nearly four hundred women who participated in the survey mentioned this as their primary way of coping. There are, I think, several good reasons. One: there is still a stigma surrounding treatment. Two: really good clinical practitioners can be hard to find; many psychiatrists these days, constrained by the low reimbursement rates of various insurance plans, basically operate prescription mills, whether or not they want to. Three: treatment for mental illness is far more expensive than treatment for other ailments.

The experience of Andra Baeten, the divorced mother we met back in chapter four, is typical. While she was still on her ex-husband's insurance, she discovered that she was entitled to eight visits with a counselor at the employee assistance program at his office. After eight visits, that counselor could refer her to an insurer-approved counselor, but the benefits brochure did not say whether this would be a psychiatrist, a psychologist, or a licensed clinical social worker. In any event, it didn't matter, because Andra tried the EAP counselor and found the visits to be a waste of time. A friend who worked for a neurologist got her the name of a psychiatrist, who Andra liked and felt was helpful. But the doctor was an out-of-network provider, so her husband's health insurance covered only half of her fee. "I quit because it got too expensive," Andra said, and she was speaking as a member of the middle class, with access to health insurance.

Merely finding the names of in-network psychiatrists can be difficult. In my insurance company's 2004 directory of preferred providers, doctors are listed by county and by specialty. If I look under my county, the specialties go from "podiatry" to "pulmonary diseases"—there's no "psychiatrist" listing at all. That's because mental health providers are listed separately, in the back, squeezed in among the listings for chiropractors, physical therapists, medical labs, and places where you can rent bedside commodes.

Some insurers limit coverage by idiosyncratic definitions: "family therapy" might be covered, but "marriage counseling" or "life coaching" may not be, even though the counseling service provided is basically the same in all cases. Other insurers limit the number of psychiatric visits a patient can receive in a given year, limits not typically placed on other kinds of medical treatment. In recent years, many companies have trimmed costs further by cutting back on coverage of prescription drugs. Not coincidentally, the newer (and often more effective) antidepressant drugs are some of the most expensive on the market. A three-month supply of Effexor XR

(150 milligram dose), for example, at the time of this writing costs at least $250.

And these are the obstacles faced by those fortunate enough to have health insurance. For those without, the options boil down to finding a psychiatrist who is willing to accept patients on a sliding fee scale (which many do), or community-based mental health programs, if they exist in your community.

Years ago, I had a recurring nightmare. It's a common one, I'm told, but in it I was being pursued, and whatever was pursuing me was something I couldn't outrun. It was always smarter than I; it knew what I was thinking before *I* knew. As time passed, my pursuer began to reveal himself. Once it was the Mud Man, a short little person made entirely of dirt the color of baby shit, who followed me around in a restaurant. As I reviled him and told him to go away, the other diners looked on incuriously; the Mud Man just stood there, smirking. Another time, my unconscious dredged up the image of an obsequious old black woman dressed in rags, who knocked at my door pretending to be starving and needy even though I knew she had come to my house to hurt me. I slammed the door in her face but, like the Mud Man, she wouldn't go away. The most vivid image, though, was a dream in which I was waiting for the doctor in an examination room. As I waited, I looked down and discovered that one of my hands and part of my arm were sprouting black, lizardlike scales, an outward manifestation of the ugliness I felt inside. I felt faintly nauseated for days.

These days, my depression reveals itself to me not in dreams, but in their absence—and in exhaustion, in the kind of irritability that makes me snarl at the telephone that has the temerity to ring, in wanting nothing more than to go to sleep forever, in fantasies of running away from the people I love more than any other people on earth.

Of all the coping methods cited by women in our survey, the one that was way down on the list—medication—is the one I used most. Maybe this is because, at a crucial point, getting on the right medication probably saved my life. I hate taking my pills, I hate the side effects (when those antidepressant ads on TV mutter about "possible changes in libido," they mean "numb from the waist down"), but there are ways around the side effects. The fact is, if it weren't for antidepressants, I probably would not be here. When I pack for a trip, my pills are the first thing I pack; when I'm at home, David has programmed my computer to boot up by asking me, "Tracy, have you taken your pills today?" At the same time, as the years pass I am more and more aware that pills are just part of the picture.

Depression makes some people lose their belief in God; it sends others on a spiritual journey. For me, it's done both. On the day I checked into the psychiatric ward fifteen years ago, I remember sitting on my bed, looking at the bare trees across the road and thinking, *If there is a God in this universe, it is not anyone I have ever met.* I was right. At that point I had not been a churchgoer for years, but I had managed to hang on to many of the beliefs I'd been taught in Sunday school, the chief one being an image of God as a kind of Cosmic Handyman. "Take it to the Lord in prayer" was the advice I'd been given, and I had, and time after time God had failed to bail me out of the mess I was in. At that low moment, I realized I'd been praying only to my concept of God; judging from the evidence, I'd been talking to myself.

For years after that day, I called myself an atheist. And then, after a long time, I began to think about whether there might be another possibility, one that had nothing to do with dogma of any kind, something beyond my power to imagine but made no less real by my limitations. The thought entered my mind just after Rebecca's birth (another gift she's given me) when, feeling the descent of that unspeakable mental darkness, I closed my eyes and said to the cosmos, "Help." I had no idea who I was talking to, or if anyone was listening; I did not know what I was asking for. I just knew that something really terrible was happening. In that moment of complete surrender, I extended my hand and something, or someone, took it. In the black chaos of that moment, I felt a moment of pure peace.

It was fleeting, but it was something I wanted again. Over the next few years, I began to notice that there were times when, feeling exhausted and defeated by some problem, I could close my eyes, abandon my own ideas of how things should be, and simply ask the Outside, "Help me see." Something about doing that subtly changed my perception, although not always immediately, and when that happened, everything else in the universe seemed to shift an inch. The peace that I felt in those moments, the sense of being profoundly understood and loved, was something I was willing to give up all of my preconceived notions to find again. Getting to that point was a struggle, and it still is. Would I have ever discovered this without the desperation that comes from dealing with a serious illness? I don't know. All I know is that this is where I am.

I've also experimented with alternative therapies. My psychiatrist, Michael Diamond, is a burly, bearded man who would look perfectly at home in a commune wearing flip-flops and love beads. He has a special interest in alternative therapies, and over the years he has suggested a whole range to me: Tai Chi, massage, SAM-e, acupuncture, Chi Gong, yoga, hypnosis. Some I've tried, some I have laughed at, some have shown promise but were impractical, and some have really worked. He suggested acupuncture for years before I overcame my knee-jerk skepticism, and it has been, by far, the most effective nondrug relief for anxiety I've found. Nobody can quite explain how acupuncture works in terms known to Western medicine, but the insertion of tiny needles into the skin at various points of the body seems to activate endorphins, the body's own painkilling biochemical, and also seems to affect the nervous system's regulation of neurotransmitters and neurohormones. All I know is that a session of acupuncture leaves me feeling as if I am floating off the table, and the relaxation effect persists for days.

Alternative therapies are the Wild West of the medical frontier these days, and you explore at your own risk. I heard for years that SAM-e could be very effective in treating depression, but it was so expensive that I was taking only one capsule (400 milligrams) a day. I was ready to write it off when my doctor talked me into tripling the dose. That did have a mood-lifting effect, but it also exacerbated my anxiety. For me, not worth it. On the other hand, fish oil capsules, which have shown promise in countering the effects of Alzheimer's and cardiovascular disease, were not only quickly effective in enhancing the effectiveness of my antidepressants; they also helped control the chronic eczema I have on my fingers. I would never have figured this out, however, if I had not been keeping a "mood log"—a day-by-day accounting of my general mental state, drug intake, external stresses, and any other assorted ailments.

The government and medical establishment have only recently begun to put research dollars into looking at the effectiveness of these alternative therapies, and for the drug companies there's vastly more money to be made patenting the next blockbuster drug than in manufacturing, say, SAM-e. Yet herbs and dietary supplements may well contain some medicinal value. The best advice I can give is to regard yourself as your own research subject. It will take time, money, and careful experimentation, but sooner or later something is likely to help.

Obstacles to Coping

For many of the mothers I heard from, the biggest day-to-day problem in dealing with depression was the lack of support they got from their families and the difficulties they had in figuring out how to take care of their own needs. "Before I had children, I would take a lot of time for myself to work things out when I began to feel depressed. I don't have that luxury anymore," one mother wrote, and another described "feeling lost, lonely, desperate, wishing for someone to see my struggle and help me, but too proud to ask for help."

Others spoke of the need to put up a front. "I had tricks I would use to help me get through the days," wrote one. "I would say positive affirmations, read self-help books, listen to positive thinking tapes. [I would] tell myself to 'smile, dammit, smile' and pretend to be happy." Another wrote, "I felt like an outside observer and as if I had to fake being happy, excited—whatever emotion was expected of me. Also [I felt I had to make] sure I didn't seem anxious or cry in front of my kids." Still others spoke of the constant, nagging sense of lost time—of wanting to feel better but not being able to, realizing that these years with their children would not last forever. "The hardest part was knowing that I was wasting time," one mother wrote. "I was not enjoying them when I was with them." In other words: guilt.

Guilt combined with depression is a vicious whipsaw: as one mother put it, "the overwhelming guilt I feel for creating the atmosphere in which my children have been raised . . . both [causes] my depression and it feeds my depression." The mothers I talked to expressed guilt over a variety of things, but they were virtually unanimous in feeling guilty about *something*. Most frequently, it was the feeling that they were setting a bad example for their children. "When Bob and I decided to have a family, my first, and what I thought was the most important, commitment to this child was like the doctor's oath—'First, do no harm,' " one mother wrote. The hardest part for her, she continued, was that depression eroded her own self-esteem, making it that much harder "to raise a child with self-esteem, a positive outlook on their life, and a solid belief that they are loved unconditionally."

"I know I should be happy and enjoying my son and I just

can't pull myself out of the depression," another mother wrote. "I wish I could be a normal parent and be active and fun." She was not alone. "There are times that every fiber of my being wants to sit on the floor and cry, rock back and forth and say, 'I QUIT.' Of course, this is not possible when you have these little people who count on me. Sometimes, I look at them and wonder why the heck they even *like* me, much less love me," another mother wrote.

Others singled out husbands as the focus of their guilt— either because they felt absent in their marriage or because their husbands were forced to take on extra household work. One mother wrote, "The hardest part for me about coping with depression and functioning as a parent is being married. I have no sexual desire whatsoever. When [we] do have sex, it lasts two minutes and there is no passion. I feel so incredibly guilty all of the time because for some reason, the attraction to my husband is out the window since I have had my daughter. He is the nicest guy in the world, and I am disrespectful to him a lot. But he still sticks around."

I am a professional in the guilt department. In my worst moments, I feel guilty about (a) the example I'm setting for my daughters; (b) the fact that David sometimes has to take up the slack around the house (and even though I return the favor when he's on one of his frequent business trips, that never seems to assuage my guilt); (c) the fact that depression or medications or both often dampen my libido; (d) that I'm not doing as much work as I "should" be; and (e) everything else. In my saner moments, I realize that there is a crucial distinction to be made between *guilt* and *regret*. *Guilt* implies complicity; it carries the connotation that you had some choice in whatever it is you feel guilt about. Nobody *chooses* to suffer from depression, so there goes that rationale. Guilt is also a narcissistic emotion; it says, "Look at how bad *I* feel about what's happening to you!"—although this is an aspect people rarely talk about. Guilt serves only one useful purpose, which is to spur growth and change. If guilt is not doing that for you, then you are indulging in something called "wallowing," which is second only to self-loathing as a waste of psychic energy.

Regret, though, is different. It's a word that doesn't get much use these days, except in the kind of nonapology apologies our culture is famous for ("I regret if anyone took offense at my racist remarks"). But the real meaning of *regret* encompasses the notion of empathy, and a desire to do whatever is possible to make amends. Do I regret struggling with depression and the impact that it has had on my children and husband? Once a day, every day, all day long. But as long as I am doing everything possible to take care of myself, and them, I know I have nothing to feel guilty about.

"I hate the fact that I am a limited human being, but I am a limited human being," Emily told me that spring day. "And there's a limit to how far I can go before I crash."

The past winter has taught her a lot, she said. Knowing that her tendency during times of depression has been to isolate herself, she has told friends, "If you don't hear from me, call me." Ideally, it should be a small circle of friends, she said, so no one person feels responsible. "You can watch out for them, and they can watch out for you." She and her husband have learned to do this for each other, too, she said. "And not in a parental way—it's more like, 'I'm worried about you.' "

Is exercise helpful? I asked. "Yeah," she said, and sighed. "Yeah, you should do it. Yeah, if you can"—spoken in the same tone she might have used to describe her failure to make brownies from scratch for the school bake sale. When her exercise program fell by the wayside, she tried to give herself a break. "I tried to carve my expectations down. I tried to turn off my Inner Perfectionist, or just say, 'I'm not going to listen to you.' " Having a running buddy was helpful, she said, because their twice-a-week schedule was often the only times she exercised—and she knew that her friend would show up and would not easily take no for an answer. "A lot of times, if I didn't have that on my calendar, I wouldn't be running."

Like me, Emily has trouble drawing closer to her children during times of depression (and, like me, she is apt to mentally beat herself up for this). What helps, she said, was making sure that her children had plenty of playdates. "It helps me realize that other parents do what I do, and that my parenting is *not* that bad."

There were days when she had to cut the day into segments, and sometimes segments into hours, and sometimes hours into single breaths. During those times, she lived between therapy appointments, which acted as a method of psychic "detoxing," she said. At first the relief came only while she was in her therapist's office. Then the effect lasted for a few minutes after she left. Then it lasted for an hour, and then several hours—and, slowly, her recovery gained momentum. Staying afloat meant sometimes casting away anything that wasn't absolutely essential, the same way boats batten down for storms. She stopped listening to National Public Radio. If the television was on with upsetting news of earthquakes or a child abuse case or war, she turned it off. She used music—sometimes quiet music to foster serenity, sometimes raucous "blast the demons out" rock music.

She learned, too, to surround herself with beautiful things all the time—an unusual rock, a drawing one of her kids had made, a cheap bouquet from Safeway—"just little things that remind me of the beauty of the world." Sunshine. The warmth of her little boy's body against hers, the silkiness of her little girl's hair. Each little thing was a foothold on that long, slippery climb out of the Pit, where even the smallest purchase makes a difference; every small piece of information was something to be stored away for future use. Next time, if there is a next time, maybe it will be easier.

CHAPTER EIGHT

How the Struggle with Depression Can Make You a Better Mother

There were all these lessons I had to learn, and I couldn't move on until I learned them, and they were just put in my path. If you don't believe, you'll call it coincidence. . . . But every time I was able to help someone, they would be helping me.

—DENISE REID, FORTY-EIGHT, OF FARIBAULT, MINNESOTA,
  MOTHER OF THREE

There is a persistent romantic view that suffering from depression gives its victims some extraordinary spiritual insight or artistic genius or tragic fragility. It's the unspoken subtext of the lives of many famous people, from Marilyn Monroe to Kurt Cobain, and less famous people have used it to rationalize all kinds of bad behavior. "Depression, in our culture," Peter Kramer writes, "is what tuberculosis was 100 years ago: illness that signifies refinement." I used to believe this, too, but as the years have passed I've come to believe it's just not true. Depression is an illness, no more or less ennobling than diabetes or hypertension or paraplegia. Depression does not inspire creativity; it kills it. When I am suffering through one of my bad spells, the effect I see my illness having on my family is the worst torture, but coming in a close second is my inability to work.

But to *engage* in the struggle with depression, to confront it and learn from it—*that,* I submit, does confer special insights. You can argue, of course, that people who deal with depression are not unique this way—that cancer and multiple sclerosis and losing one's family in a terrorist explosion all qualify as searing, life-altering events, which is true. But when you are surveying the wreckage of your life—and the exact cause of the devastation at some point ceases to be relevant—the question you must confront is the same one depression poses: *How do I live now?* The unique nature of depression is that it gets you to that question directly; no external catastrophe is required.

The difference between mere suffering and the sense we make of our suffering is a subtle one, but it was succinctly put by writer Joshua Wolf Shenk in his recent book about Abraham Lincoln, who historians agree was a victim of recurrent and severe depression. Lincoln's greatness didn't come from simply suffering depression, or from some miraculous triumph over it, Shenk writes, because the illness plagued him throughout life. "Rather, [Lincoln's greatness] must be accounted an outgrowth of the same system that produced the suffering." Instinctively, I think most people sense that there is an important distinction here, although our frequent confusion about it explains the persistent fear that antidepressants are just "happy pills" that will turn everyone into bland, cheerful little robots.

The qualities that Lincoln developed in his struggles with depression helped him guide the nation through the Civil War, and in my own small way I can make a similar claim: I am a better mother for having had to deal with this illness. It's taught me that I am tough: I am easily knocked down, but eventually I get up again—and that is a quality I hope my girls can learn from. It has taught me hope: in times of great darkness, something gets me through. It's taught me that I can make fundamental changes in my life when I have to. It's taught me self-respect: even though I have thought of myself as worthless too many times to count, there have been crucial moments when other people treated me as a person of great value—not because of anything I did or had, but just because I was me. Eventually, I learned to think of myself that way. Now, every day, I try to convey the same message to my daughters: *You are already perfect.* Finally, my struggle with depression has triggered in me a strong need to learn and understand it—and to pass on that knowledge to other people who might need it.

Most important, it has taught me the kind of empathy that I don't think I would have developed any other way. I did not correctly diagnose Rebecca's pain right off the bat—I don't think many parents would immediately recognize signs of emerging depression in a child so young—but I did know from the beginning that we were dealing with more than a discipline problem. That unwavering conviction did, I think, keep me and my husband from traveling down a lot of blind alleys and forestalled the escalating and destructive series of power struggles that would have happened had we labeled her behavior as simply acting out.

How does empathy develop? Through self-awareness. Re-covering from depression requires a lot of self-awareness—specifically, the ability to recognize your own moods, what triggers them, and how to manage them. Nearly half of the mothers I talked to or heard from said that empathy was the skill that surviving depression helped them develop, and not just empathy for their children, but for all kinds of people. "[Depression] has allowed me not to place judgment on others, and has taught me that unless you have walked in that person's shoes, you have no idea what they are going through or have gone through," one mother told me. "I say this because I was often treated as if I were bad or lazy, when others had no idea of the terror that was going on in my mind while they judged me."

Just as real courage isn't possible without suffering fear, the development of empathy isn't possible without having suffered the acute sense of your own imperfections. Pity looks down; empathy sees eye to eye. As another mother put it, "I believe that my depression has given me more humility and acceptance. Through the process of healing, I have undergone a spiritual healing. . . . This perspective gives me more empathy toward other people that I meet."

There were hints of a sea change in societal attitudes toward depression in the responses, too. Many mothers said the single greatest benefit of their own experiences was that it taught their children to recognize the symptoms of mental illness, to not feel ashamed of it, and to seek help when they needed it. As one mother put it, "I've tried desperately hard to learn from my mistakes, and improve, and show [my daughters] that there's no such thing as being perfect."

"We have been able to be open with one another and not live in denial," another mother wrote. "I have been able to point out areas where I think they are following in my path and we can at least talk. Neither allows [himself] to be manipulated by me and [they] don't give in to me when I'm being selfish and withdrawn. Often they are the mirror in which I see myself, and they motivate me to do more." Said another: "It has made me acutely aware of my own history, and helped me to stop the generational [cycle of] abuse. . . . My children have never experienced abuse from me."

"Depression has forced me to look at my upbringing and the effect it has had on me," another mother replied. "I am [now] acutely aware of how my behavior affects my children, and sometimes I can actually stop myself midreaction and change course." Wrote another: "The techniques I have learned in therapy for my depression help in everyday situations and I can teach my children to use these techniques in dealing with their problems, even at the ages of eight and three."

Self-awareness is a skill—and as some of these mothers have learned it, they have tried to pass it along to their own children. "I feel I've been a good role model for expressing emotions and physical feelings. 'I'm so tired after this busy day.' 'I'm so happy we found you a good cello teacher.' 'Doesn't it feel good to have all your math done?' " wrote one mother. She also thinks her experiences have shown her children the value of respecting their bodies and the subtle difference between vanity and self-care. "Over the past few years they've noticed my manicures, pedicures, putting on makeup (including lipstick) most mornings, and opening the doors and windows to enjoy the fresh air and sunshine on a beautiful day," she added. "Now they see me taking piano lessons and working out at the fitness center most every day."

Another mother who answered this question works as an office manager for a psychiatrist. One day, she said, a patient who was unaware of her struggle with depression asked her what it was like, being around "crazy" people all day. "I told her, I don't think of our patients that way at all, but as people working so very hard to get and be well, and how I applaud her and the others and admire their tenacity." Her own daughters, she said, "have seen me work hard to get well and maintain that state. They see this as an ongoing struggle, not just something that hours in talk therapy or the right meds can miraculously [cure]."

"I think seeing me battle the depression, its severity, and my eventual success in beating it . . . have shown my children the value of perseverance," another mother replied. "It's also shown me that taking care of myself is taking care of my children. If I hadn't taken care of myself, they would be motherless [or] I would still be using negative parenting techniques. The depression forced me to get help for all of us."

Then there is courage. "Because I have had to dig myself out of that dark hole so many times, I think depression has given me the self-confidence to know I can defeat great obstacles," one mother wrote. "I think I share that self-confidence with my children and give them the notion that with hard work you can accomplish anything."

The biggest benefit for a few women, though, was that somehow, surviving depression taught them the skill I've always found so difficult: simply being in the moment, learning not to sweat the small stuff. "I used to be a micromanaging mom at home: my husband couldn't do the laundry/dishes/house-cleaning correctly, I couldn't cut myself any slack and was obsessive, almost, about following the routines of the house to the nth degree," one mother wrote. But having to depend on her husband during a period of acute depression taught her that her husband could handle the housework just fine, and so could her kids, "which builds up their self-esteem in the long run. I'm a better mother now having gone through my depression, because I realize now the work of the home does not ever matter—it'll never be done, not with four kids here! What does matter is how well I've cared for them over the course of the day. They will not remember the Pottery Barn–perfect home they could have lived in, but the fun and happiness that was a part of each day."

Depression is an ugly illness that can sometimes reveal beauty, even if it's just the beauty of our own plain, unadorned selves. I remember the mother of a friend of mine in child-hood—a woman whose every comment was painfully correct, whose outfits were always impeccably pressed and neat, whose house smelled like Lysol. Even as a kid, I knew she was a tortured person. I never wanted to be anything remotely like that—but even so, it's taken me a big chunk of my life to learn that I don't have to pretend to be anybody other than who I am. After years of hiding my face under layers of makeup and my tortured psyche behind a facade of achieve-ments, plain is beautiful to me now. Emotionally speaking, I'm still a work in progress. I do not always take good care of myself. Sometimes I allow my casual attitude toward personal hygiene to slide into downright sloth. I eat unwisely (especially under stress); I do not get enough exercise. Sometimes I yell at my kids and neglect my husband. But I am not twisting myself into knots to be someone I am not—and somehow, for whatever reason, my children and my husband love me anyway, and I love them right back.

The fierce arguments David and I have had over dealing with Rebecca have disabused me of any idea I may have once entertained that I had married a mellow, laid-back model of patience. On the other hand, what we've been through together has shown me that I married a man who will not give up, who fights fair, and who is not willing to let my illness come between the two of us. Whenever depression threatens to erect a wall between us, sooner or later he drags me back, kicking and screaming. Not only does he love the real me, he gets pissed off when he doesn't get the real me. And, when I think about it, "authenticity" is about as good a one-word definition of emotional health as anything else I can come up with.

"I am real now," one mother writes. "Not hiding behind a happy face."

Denise Reid is forty-eight, a secretary in a school for the deaf in Faribault, Minnesota, about an hour south of Minneapolis. She has been married for twenty-nine years to Jim Reid, a middle-school teacher. A plump woman whose blond hair is cut in a short, no-nonsense style, she radiates a friendly sense of calm and laughs frequently.

She had no experience with depression until she had her first daughter, Jennifer, at twenty-two, and experienced postpartum depression. That was 1978. She white-knuckled her way through that, and four years later gave birth to another daughter, Heather, and again suffered PPD. Then, in 1986, when her daughters were four and eight, she fell into the worst depression she had experienced so far. Up to that point, she said, "I could do anything and everything." She worked full-time, was a wife, mother, household manager; everything ran the way it was supposed to. She was a superorganized Supermom. "I even sewed my husband's clothes. I could just do *everything*." And then, suddenly, "I couldn't do anything. So there was no worth in me anymore. And who is going to want me around?"

She did not try medication at that point, but did find a therapist she could work with, and slowly began to feel better. By 1989, she was herself again; she remembers Christmas of that year being especially happy, and thinking gratefully that all that work in therapy had really paid off. Then, in 1990, she got pregnant with her third daughter, Danielle, and the bottom fell out again. This time, she was hospitalized.

In the hospital, there were no decisions to be made, no chores to be done, no newborn to take care of. Her job was simply to heal. She used the time to read, and what she began reading were books about the biochemistry of depression. "It was a real turning point," she said. For the first time, she really allowed herself to believe that her mental pain was not imaginary—that there were biochemical abnormalities in her brain that were making her ill. It was a revelation. After that, she said, she gained the courage to talk about her illness, to reveal a side of herself to friends and neighbors that she had never shown before. One thing led to another. Some months later, she was picking Danielle up one day from her neighbor's house when she saw another mother at the doorstep, looking teary-eyed. "This is Lisa," her day care provider said. "I think she has what you had."

Denise laughed. "And then *she* gets hospitalized, and then we founded a support group of six of us. We just latched on to one another."

Denise thinks that people like herself—who are blessed with natural confidence, who when things are going well are particularly capable and efficient—have the most to learn from depression. "It rips everything from you: your self-worth, your self-confidence. You're laid bare. There's nothing you have to give anybody." It was only then, when it was no longer possible to define herself by what she could *do*, that she learned that it was okay to just be. This did not come as a dramatic epiphany, but as a series of small steps, most of which seemed inconsequential at the time. She got up the nerve to talk to a friend about her illness in an honest way, and that friend did not run away. She kept getting up in the morning even when she didn't feel like it. She adopted a mantra: "I will accept the situation as best I can and allow myself happiness in spite of difficulty." She learned that she could not safely drink. She learned that sleep is "a powerful tool" for getting better. Every small thing built on what went before.

As she spoke, she sat in an easy chair in the small den of her house; while we had been talking, her youngest daughter, Danielle, had come home from school. Danielle is fourteen—a leggy young girl with luxuriant honey-brown hair and braces, perched on that elusive boundary between childhood and womanhood. Unselfconsciously, she came into the room where we were and plopped down into her mother's lap. Supposedly, she was there to ask a question—something about an allowance or a movie outing—but I could tell she was listening.

"It was a gradual process," Denise was saying. "But at some point I looked back and I realized I had known it for a while." And at the end was the realization that "I know my self-worth now. It's not what I can do or give. I know I am worthy, just because I am a human being." There were plenty of times when she wasn't sure she could make it; there were plenty of times when all three of her daughters, as well as her husband, suffered the collateral effects of her illness. She credits her survival to the three F's—faith, friends, and Finnish genes—and also, in no small part, to her children. "There was a *reason* for me to push and try."

Denise lives on a quiet street of small, immaculately kept homes. Her house sits on a rise, set off from the street by a flight of concrete steps. As I am leaving, she walks out with me and stands at the top of those steps to say good-bye. The last I see of her, she is smiling, raising her hand to wave. For a moment, the wave looks like a benediction.

The story that has always seemed to epitomize the struggle with depression is the Greek myth of the Minotaur, a horrifying beast with the body of a man and the head of a bull. The product of an illicit liaison between a magical white bull and Pasiphaë, queen of the kingdom of Crete, the Minotaur was imprisoned by the queen's husband, King Minos, in an enormous labyrinth. To pacify the beast, King Minos periodically put human sacrifices into the labyrinth, where they would wander, terrified and lost, until the Minotaur found them and ate them. The Minotaur was finally slain by Theseus, who entered the labyrinth with a ball of string and who, by unspooling the string as he went, left himself a trail to follow to get back outside after the battle.

I suppose I heard that story in eighth grade or so, and the myth conflated itself in my mind with those recurring dreams I had about being pursued. It was as if I'd been there. I could feel the dread of those human sacrifices; I seemed to remember the deadly quiet inside the labyrinth; I could imagine my own heart pounding so loud it just had to be audible; I felt, as if it had happened to me, the paralyzing terror of hearing a soft footstep in the dark, just a few feet away. The need to escape must have been overpowering—and yet the smallest movement could tell your enemy where you were. At any turn you might come face-to-face with another person as lost and terrified as you, or with the beast himself; you never knew. And yet you had to know. You had to move. Not to move was death.

In my life, though, nobody killed the beast. Instead, I met the beast, and the beast was me, and somehow or other I found that thread that led me out of the labyrinth. I've been in and out of the labyrinth many times now, and even when I feel lost inside it I know that if I hang in there, somewhere there is a thread; there is a way out.

My mother never knew that. She spent most of her life hiding from the beast, who stalked her to the end. Stroke in itself is notorious for causing depression.

"What's your book about?" she asked me one day, for what seemed the twentieth time, and I said, "It's about motherhood and depression."

"Oh," she sighed. "Well, I could tell you a thing or two about that."

"I know you could, Ma. I know," I said, thinking: *You already have.* Now that I am a mother myself, I have jettisoned my old grievances against her. There are conversations we will never have now, and maybe that's for the best. In fact, I honor her for the immense effort she put into being the best mother she could, despite the heavy emotional baggage she carried. Had I been in her situation, I could not have done any better. I probably would not have done as well.

She slept a lot those days, living her life more vividly in her dreams than in reality, the line between past and present becoming increasingly blurred. In the fall of 2004, she had one of her periodic health crises—she suffered from congestive heart failure, and even a slight cold could put her at death's door—and I went down to Atlanta half expecting that that time she wouldn't make it. But she did. I sat with her until late that night, and then went to bed myself, and in the morning she was much better. I sat down again in the easy chair beside her bed. It was midmorning; I had closed the blinds against the glare of the late November sun.

Suddenly my mother awoke with a start and turned toward the window. She squinted uncomprehendingly at me for a full moment. Then she said, "Oh, it's you. I couldn't see who you were with that glare behind you. I thought you were an angel."

"It's just my natural radiance," I said lightly. "You know, from my angelic nature." She chuckled comfortably and I reached for her hand. We sat there a minute.

"I love you, Ma."

"I love you too, darlin'," she said, and then she added something she had never said to me before, not ever. "I love you just the way you are." And just for a moment, with her hand in mine, it was as if nothing bad had ever happened to either of us.

Introduction

xvi **I wrote a memoir of my experience of coming to terms with this illness:** *The Beast: A Reckoning with Depression* (Putnam, 1995), reprinted in paperback under the title *The Beast: A Journey Through Depression* (Plume, 1996).

Chapter One: What Is Maternal Depression?

3 **women make up roughly 12 million of the 19 million Americans affected by depression every year:** National Institute of Mental Health (NIMH), unpublished epidemiological cachement area analyses (1999).

3 **About one woman in every eight can expect to develop clinical depression during her lifetime:** NIMH, unpublished epidemiological cachement area analyses.

3 **the incidence of depression in women peaks between the ages of twenty-five and forty-four:** NIMH, "Depression: Treat It. Defeat It." Accessed June 1999.

3 **assume . . . roughly one-third of the 12 million women currently affected by depression . . . have children at home:** An estimate I took from Bureau of Labor Statistics data. "Work at Home" (May 2001) says that of 62.1 million women who worked at home that year, 39.7 percent had children under

223

eighteen. "Women in the Labor Force" (February 2004) says that 32.1 percent of women in the civilian nonindustrial population have children under eighteen; of those actually in the labor force, 38.7 percent had children under eighteen. This estimate assumes, of course, that women with depression are more or less representative of the population as a whole. However, given the underreported nature of depression, my estimate may be even more conservative than it looks. In another attempt to estimate the scope of the problem, Dr. Myrna Weissman of Columbia University took a sample of 345 mothers with children under the age of eighteen who were coming to see a primary care doctor for routine medical care. She found that 25 percent of the mothers were currently depressed. (Myrna Weissman and Peter Jensen, "What Research Suggests for Depressed Women with Children," *Journal of Clinical Psychiatry* 63 [July 2002]: 7.) Extrapolating that percentage according to the 2000 U.S. census: there are 209.1 million adults in the United States, 104.5 million of whom are women. If one-third of those women have children under eighteen, that comes to roughly 35 million women; 25 percent of 35 million would equal 8.7 million mothers currently suffering from depression. This figure, of course, is my conjecture. The point is, nobody knows the number. As Weissman points out, women of childbearing age have historically been excluded from the bulk of research involving depression, that is, studies involving depression drugs, out of concern for the potential effects on fetuses.

4 **said they considered depression a normal part of aging, menopause, and the postpartum period:** National Mental Health Association, "American Attitudes About Clinical Depression and Its Treatment" (March 27, 1996).

4 **fewer than half the women who experience depression ever seek medical help:** A. Rupp, E. Gause, and D. Regier, "Research Policy Implications of Cost-of-Illness Studies for Mental Disorders," *British Journal of Psychiatry* 36 (1998 Suppl): 19–25.

6   **largely reflected the demographics of O: *The Oprah Magazine* readers:** I should also note that the number of women answering each question varied considerably; but questions about demographic status had by far the highest response rates. Also, the women who responded via my newspaper inquiries tended to be older and/or less affluent than the women who responded via O: *The Oprah Magazine*.

7   **Sherryl and I used some of those tools, too, in order to measure the women's responses against some kind of objective yardstick:** We included a twenty-question survey known as the Centers for Epidemiological Studies–Depression Inventory, widely considered a reliable screening tool for evidence of depressive symptoms. Of the 393 women in the study, 327 provided answers to all twenty questions. Scores could range from 0 to 60. Of those, 11 women scored under 16, indicating they were not at risk for depression. Thirteen scored between 16 and 20, indicating the presence of mild depressive symptoms. The rest—303 women—had scores of 21 or more, indicating moderate to severe depression, and of the last group, 46 had scores of 40 or more, indicating severe depression. The survey was available online from April through August 2004. Each respondent could take it only once, so these numbers are only a snapshot indication of the respondents' emotional well-being.

11   **the pioneering work of Myrna Weissman and Eugene S. Paykel in the early 1970s:** Myrna Weissman and Eugene Paykel, *The Depressed Woman: A Study of Social Relationships* (University of Chicago Press, 1974).

11   **the cutting-edge research in this field:** A complete list of leading researchers in this field is too long for a footnote, but a few of the institutions in the forefront in this area are Harvard University, notably Lee Cohen; the University of California, Los Angeles, notably Lori Altschuler; and Emory University, notably Charles Nemeroff, Zachary Stowe, and Paul Plotsky. Dr. Sherryl Goodman, who designed the survey that is the

foundation of this book, is coeditor, along with Ian Gotlib at Stanford University, of one of the leading texts in the field, *Children of Depressed Parents: Mechanisms of Risks and Implications for Treatment,* published by the American Psychological Association (2002). Tiffany Field at the University of Miami is also a noted expert on the effects of depression on neonates, as is Vivette Glover of Queen Charlotte's Hospital in London. Myrna Weissman, of the New York State Psychiatric Institute, is now in the fourth generation of a study tracking the effects of maternal depression. Daniel Pine at the National Institute of Mental Health in Bethesda, Maryland, is doing notable work on the treatment of anxiety in children, which is a frequent marker for an increased vulnerability to depression in later life.

15　**suffer depression at more than double the rate of people who have another type:** Avshalom Caspi et al., "Influence of Life Stress on Depression: Moderation by a Polymorphism in the 5-HTT Gene," *Science* 301 (July 18, 2003): 386–89.

Chapter Two: Motherhood

19　**"Who would want to live feeling that way?":** Linda Gray Sexton, *Searching for Mercy Street: My Journey Back to My Mother* (Little, Brown, 1994), 11.

20　**administrative assistant and managerial jobs rank numbers one and two, respectively, in terms of mental stress:** In 2004, the segment of the salaried workforce reporting the second highest number of days away from work due to anxiety, stress, or other psychological disorder were people in managerial work or in professional specialties. Data from National Institute for Occupational Safety and Health, *Worker Health Chartbook* (2004).

20　**depression is fundamentally a dysfunction of the body's reaction to stress:** A scientific overview of this connection can be found in Ronald Duman, "The Neurochemistry of Depressive Disorders" in *Neurobiology of Mental Illness,* ed. Dennis S. Charney, Eric J. Nestler (Oxford University Press, 2004), 421. For those interested in a layman's explanation, written by one of the field's leading researchers, a good source is Robert Sapolsky, *The Trouble with Testosterone: And Other Essays on the Biology of the Human Predicament* (Scribner's, 1997).

21 **"intensive mothering":** Sharon Hays, *The Cultural Contradictions of Motherhood* (Yale University Press, 1998).

21 **The bar has been raised in imperceptible increments:** An excellent and detailed discussion of how this happened can be found in Susan J. Douglas and Meredith W. Michaels, *The Mommy Myth: The Idealization of Motherhood and How It Has Undermined Women* (Free Press, 2004).

23 **a mother's vigilant watchfulness for the child's every waking moment:** For an excellent deconstruction of how these assumptions are woven into the most widely read child-rearing manuals, see chapter three of Hays's book *The Cultural Contradictions of Motherhood.*

24 **the belief that "there is something . . . children need that only [mothers] can give them":** Ann Oakley, *Woman's Work: The Housewife Past and Present* (Random House, 1976), 211.

27 **"barely recognize amid the passion either the thinking or the work":** Sara Ruddick, *Maternal Thinking: Towards a Politics of Peace* (Ballantine Books, 1989), 67.

28 **"I often wondered why mothers didn't get tired":** *Voices of American Housewives,* ed. Eleanor Arnold (Indiana University Press, 1985), 115.

31 **"natural or practical intelligence, wit, or sense":** *The Random House College Dictionary* (1975).

32   **has been directly linked to major depression:** Angela L.
Lee, William O. Ogle, and Robert M. Sapolsky, "Stress and Depression: Possible Links to Neuron Death in the Hippocampus," *Bipolar Disorders* 4:2 (2002): 117–28.

34   **a child is roughly three times more likely to develop one
too:** Myrna Weissman, Virginia Warner, Priya Wickramaratne,
Donna Moreau, and Mark Olfson, "Offspring of Depressed Parents: 10 Years Later," *Archives of General Psychiatry* 54 (October
1997): 932.

Chapter Three: What's Wrong with Me?

44   **"Did anyone call to see her or look after her all the while
she was there?":** Files at the National Archives from St. Elizabeth's Hospital. Elizabeth Dreher, case # 4083; letter to Superintendent C. H. Nichols, dated May 14, 1877; letter to Nichols,
dated October 31, 1876; letter from John L. Dretcher to Dr.
Godding at St. Elizabeth's dated March 22, 1898; letter from
Dretcher to Godding dated April 8, 1898.

45   **having a history of depression prior to childbirth makes
PPD more likely:** Michael W. O'Hara and Laura L. Gorman,
"Can Postpartum Depression Be Predicted?" *Primary Psychiatry*
(March 2004): 43. O'Hara's conclusions were borne out by the
numbers of our survey: among the 393 women who responded,
83 percent of whom had a history of depression before becoming mothers, 58 percent said they had suffered PPD at least
once. That's roughly three times as high as the best estimates of
the incidence of PPD in the general population, which is between 10 and 20 percent.

45   **PPD is all too often left undiagnosed and untreated:** Of
the 68 women in our survey who never experienced depression
before becoming a parent, 43 said that their first experience of
depression was PPD. Nor was this a simple case of feeling bad
for a few days or weeks: of the 229 women who answered our

question about the duration of their PPD, 43 percent said that their symptoms persisted for at least one year.

45   **that can have lasting repercussions:** Myrna Weissman, Virginia Warner, Priya Wickramaratne, Donna Moreau, and Mark Olfson, "Offspring of Depressed Parents: 10 Years Later," *Archives of General Psychiatry* 54 (October 1997): 932.

45   **an episode of PPD increases the likelihood of future depressive episodes in general:** Interestingly, British researchers have found that for women with a history of depression, one episode of PPD does not increase the likelihood of future episodes of PPD specifically, though it does increase the likelihood of depressive episodes in general. For women whose first experience of depression was PPD, one episode increases the risk of future episodes of PPD, they found. Peter J. Cooper and Lynne Murray, "Course and Recurrence of Postnatal Depression," *The British Journal of Psychiatry* 166 (1995): 191–95.

46   **"other moms who seem to have it so great":** Letter from Deborah Chaney, April 30, 2004.

47   **"now if you think something is wrong with you, get a family doctor":** Letter from Meredith Goodlatte, May 20, 2004; interview August 30, 2004.

47   **"the mark could be seen for the rest of her life":** Cited by Rebecca Shannonhouse in *Out of Her Mind: Women Writing on Madness* (Random House, 2000), 3.

48   **"very subject to headaches and to madness":** Thomas Ewell, *Letters to Ladies: Detailing Important Information, Concerning Themselves and Infants* (Ewell, 1817), 237.

48   **"We made sure of that":** Letter from Wendy Morgan, May 22, 2004.

48   **"sounded like they were underwater":** Letter from Sandra
      Coleman, April 22, 2004.

48   **"watching myself without any emotion":** Letter from
      Stephanie Walters, May 10, 2004.

48   **"I hate that I lost that time with her":** Letter from Kathryn
      Brown, April 26, 2004.

49   **the rate at which they drop:** Leslie Born, Dawn Zinga, and
      Meir Steiner, "Challenges in Identifying and Diagnosing Post-
      partum Disorders," *Primary Psychiatry* 11:3 (March 2004): 30.

49   **many cases of PPD are simply a worsening of depressive
      symptoms that were present before childbirth:** Interview
      with Dr. Lori Altschuler, August 18, 2004.

50   **emergency C-section:** Born et al., "Challenges."

50   **sleep deprivation is "the root of all evil":** Interview with
      Shari Lusskin, March 25, 2004.

51   **Women have an innately greater sensitivity to noise than
      men:** Jamie L. Rhudy and Mary W. Meagher, "Noise Stress and
      Human Pain Thresholds: Divergent Effects in Men and
      Women," *The Journal of Pain* 2:1 (February 2001): 57–64.

53   **still show signs of it after one year:** Born et al., "Challenges."

53   **"unconscious resentment" toward her baby:** Letter from
      Jeannie Wegman, April 18, 2004.

58   **began to suffer unusual anxiety:** Interview with Laurel
      Spence, June 8, 2004.

61   **couldn't bring herself to admit it for months:** Interview
      with Paul Cook, M.D., June 8, 2004.

63 **"It is not something we [were] trained to ask about"**: Letter from Kirsti Dyer, April 16, 2003.

65 **"how could anybody hurt something so precious"**: Interview with Sunshine Gage, June 8, 2004.

67 **"I must be crazy"**: Private letter to author, June 28, 2004.

71 **from Zoloft, Paxil, or Luvox**: New studies, published just as this book was going to press, raised the possibility that using Paxil during pregnancy increased the risk of infant seizures, but those findings had not been duplicated.

71 **will usually relapse within a few months**: Lee S. Cohen, Ruta M. Nonacs, Jennie W. Bailey, Adele C. Viguera, Alison M. Reminick, Lori L. Altshuler, Zachary N. Stowe, and Stephen V. Faraone, "The Relapse of Depression During Pregnancy following Antidepressant Discontinuation" (unpublished; NIMH grants # 19445 and # 56420–05).

72 **"Until doctors are better educated, patients will continue to get short shrift"**: Interview with Shari Lusskin, March 25, 2004.

72 **"they didn't really have a lot of information on [antidepressants] and nursing"**: Interview with Jeneva Patterson, March 18, 2004.

73 **"They deferred to me"**: Interview with Virginia Major, March 23, 2004.

74 **and more than a little fear**: A few weeks after our conversation, Jeneva relapsed into serious depression. After reinstituting her medications, and with the help of her obstetrician, her psychiatrist, and her therapist—not to mention her husband—she recovered fairly quickly. "This battery of specialists rallied behind me and I was taken care of," she said. "The system

worked"—although, she said, she had to go through four obstetricians and four therapists before she was able to construct a
team of doctors that was right for her. Her baby, Nicholas Gor-
don, was born September 8, 2004.

Chapter Four: The Way It Is

92 **alcohol, prescription drugs, or illicit street drugs:** A total
of 330 women answered the question, "Have you ever used alcohol or street drugs or misused prescription drugs to help you
with your symptoms of depression?" Of those, 139 (42.1 percent) said yes.

95 **"I might not have to deal with it now":** Telephone conversation, March 18, 2004.

98 **simply by being put in an institution for decades:** A har-
rowing collection of some of these accounts may be found in
Jeffrey L. Geller and Maxine Harris, *Women of the Asylum:
Voices from Behind the Walls 1940–1945* (Doubleday, 1994).

100 **to set out milk and snacks for her children, for when
they woke up:** Edward Butscher, *Sylvia Plath: Method and
Madness* (Seabury Press, 1976), 352–57.

100 **"I just wanted to hide under a rock":** Interview, July 16,
2004.

100 **"a marriage of friendship, not of love":** Interview, June 1,
2004.

Chapter Five: Rats, Monkeys, and Mothers

120 **why drugs that get into the bloodstream within an hour
can often take weeks to work:** Rene Hen et al., "Require-
ment of Hippocampal Neurogenesis for the Behavioral Effects
of Antidepressants," *Science* 301 (August 8, 2003): 805.

123 **even drink too much:** "These monkeys also tend to consume excessive amounts of alcohol when placed in a 'happy hour' setting as adolescents and young adults." Stephen J. Suomi, "Gene-environment Interactions and the Neurobiology of Social Conflict," *Annals of the New York Academy of Science* 1008 (2003): 133.

123 **with the "long" version of the same gene:** Even genes that perform extremely specific functions, like transporting serotonin between nerve synapses, can come in various styles (in this case, long or short). In the Caspi study, researchers followed a group of 847 Caucasian New Zealanders from birth in the early 1970s into adulthood. As the number of stressful life events (things like job loss, relationship woes, or money trouble) increased, those people born with two copies of the short serotonin transporter gene had the highest rate of depressive episodes (a number researchers corroborated by talking to family members), and this was true whether or not those people had a history of depression. Those born with two copies of the long version had low rates of depression, no matter how many stressful life events they suffered. Avshalom Caspi et al., "Influence of Life Stress on Depression: Moderation by Polymorphism in the 5-HTT Gene," *Science* 301 (July 18, 2003), 386–89.

127 **Now I was looking:** Plotsky's thoughts on the ethics of animal research are clear: a sign in his office says ANIMAL RESEARCH SAVES LIVES! He works with rats because their relatively short life spans allow him to see in a matter of weeks the effects of heredity and environment that would not be apparent in primates for years. But he also admitted feeling ambivalent at times. Working with primates was emotionally stressful, he said, and he gave up doing it partly for that reason. I found this admission oddly likeable.

128 **a "profile of dysregulation" for the first year of life:** Tiffany Field, "Prenatal Effects of Maternal Depression," *Chil-*

dren of Depressed Parents: Mechanisms of Risk and Implications for Treatment, ed. Sherryl H. Goodman and Ian H. Gotlib (American Psychological Association, 2002).

128 **early-life stress . . . permanently increases the brain's sensitivity to stress:** Charles B. Nemeroff, "Neurobiological Consequences of Childhood Trauma," *Journal of Clinical Psychiatry* 65 (2004): 18.

128 **the hippocampus in women who suffered physical or sexual abuse as children was smaller than normal:** Meena Vythilingam et al., "Childhood Trauma Associated with Smaller Hippocampal Volume in Women with Major Depression," *American Journal of Psychiatry* 159 (December 2002): 2072.

129 **an astounding 67.6 percent already had acquired some kind of psychiatric diagnosis by the time they were at or approaching puberty:** Myrna Weissman et al., "Families at High and Low Risk for Depression: A 3-Generation Study," *Archives of General Psychiatry* 62 (January 2005): 29.

130 **The aim . . . was to get a population sample as representative of the general population as possible:** A major limitation of the study was that it excluded African Americans. At the time the study began, in the early 1970s, Kendler said, the population of Virginia was only 19 percent African American, and there were no other ethnic groups of any size. To avoid any possibility of skewing his statistics, Kendler said, he was obliged to limit his study to Caucasian twins. Interview with Kenneth Kendler, February 23, 2005.

132 **"even in the presence of high genetic risk and severe stressful life events, the majority of individuals do *not* develop an episode of major depression":** Kenneth Kendler, Ronald Kessler, Ellen E. Walters, Charles MacLean, Michael C. Neale, Andrew C. Heath, and Lindon J. Eaves, "Stressful Life Events, Genetic Liability, and Onset of an Epi-

ode of Major Depression in Women," *American Journal of Psychiatry* 152 (June 1995): 833.

Chapter Six: Don't Look Now

136 **"And add some extra just for you:"** Philip Larkin, in "This Be the Verse," *Collected Poems,* ed. Anthony Thwaite (Farrar Straus Giroux, 1988), 180.

137 **a mother who doesn't find much to like about herself is likely to raise children who will wonder if there's anything worthwhile about themselves:** Fathers, obviously, play a powerful role here, too—a role that is beyond the scope of my research. I do know that my own father's evident pleasure in being around intelligent women did a lot to buffer the other influences on me, even if he did express his feelings in sexist terms: he liked women, he told me, "who could think like a man."

138 **"I never laughed so hard in all my life":** Used by permission of Dede Walser.

140 **"and fathered a daughter by her":** Interview with Mattie Belle Buchanan, April 20, 2005. Miss Buchanan is the half sister of Enley Eugene Buchanan, and is his only surviving sibling. I should mention here that I have no definitive proof that her half brother is my grandfather, but the name is unique and the circumstances of her brother's life dovetail with the details of my mother's birth. There is also a physical resemblance. I found Miss Buchanan, as well as a picture of Enley Eugene Buchanan, through the help of Lynn Bryant, a very kind amateur genealogist and historian in Sylva, North Carolina.

141 **their mother "just cried and cried":** Interview with Dede Walser, July 5, 1988.

141 **"and that was the last I saw of her":** There's no third-party

method of verifying the story about the coat, but my aunt's memory seems trustworthy: nearly seventy-five years later, her recall of the exact time and date of her mother's death matched the information on my grandmother's Fulton County, Georgia, death certificate, which I found in the course of researching this book.

142 **"I'll be your mommy":** Interview with Ruth Thompson, July 4, 1988.

145 **"neither sorrow, nor crying, neither shall there be any more pain":** Revelation 21:4, King James Version

152 **we finally got Rebecca tested:** Rebecca was tested with the Wechsler Intelligence Scale for Children—Fourth Edition; the Wechsler Individual Achievement Test—Second Edition; Developmental Test of Visual-Motor Integration; Rey Complex Figure Test and Recognition Trial; Revised Children's Manifest Anxiety Scale; Children's Depression Inventory; Behavior Rating Inventory of Executive Functioning; Test of Variables of Attention. Finding a person who was skilled in both testing and interpretation, and who was covered by our health insurance, took me nearly a year. It is noteworthy that, without exception, the most skilled people I found were not listed in any health insurer's "preferred provider" list; they could be found only by word of mouth.

155 **most, in fact, never will:** Estimates of the rates of psychiatric disorders among children of parents with unipolar (as opposed to bipolar) depression hover around 40 percent. *Children of Depressed Parents: Mechanisms of Risk and Implications for Treatment,* ed. Sherryl H. Goodman and Ian H. Gotlib (American Psychological Association, 2002). When we asked survey participants if any of their children had ever been treated for problems with their emotions or behavior, 57 percent (118 out of 206 respondents) said no. I considered that a remarkably healthy percentage, considering that our sample was so heavily

weighted toward women who have had recurrent and serious experiences with depression.

160 **There are ways to help children who are in crisis, or who are at high risk for suffering from depression:** Seligman's book *The Optimistic Child: A Proven Program to Safeguard Children Against Depression and Build Lifelong Resilience* (Harper Perennial, 1995) is a good place to start and contains surveys and ratings that you can give to your child. Another is Ronald M. Rapee, Susan H. Spence, Vanessa Cobham, and Ann Wignall, *Helping Your Anxious Child: A Step-by-Step Guide for Parents* (New Harbinger Publications, 2000).

161 **cognitive behavioral therapy in particular has been shown to be highly effective:** If you're interested in an academic understanding of the theory behind cognitive behavioral therapy, the classic text is Aaron Beck, A. John Rush, Brian F. Shaw, and Gary Emery, *Cognitive Therapy of Depression* (Guilford Press, 1987). For an approach based on Beck but more geared to the layman, a good resource is David Burns, *Feeling Good: The New Mood Therapy* (Avon, 1999), which is sold with assorted workbooks for homework assignments. Martin Seligman's work looks at this from the flip side—what does it take to make us happy?—and his book *Authentic Happiness: Using the New Positive Psychology to Realize Your Potential for Lasting Fulfillment* (Free Press, 2002) is an effort to help define and develop specific personality strengths that can be used to ward off depression. Daniel Goleman, *Emotional Intelligence: Why It Can Matter More than IQ* (Bantam, 1995), provides an overview of this subject that is particularly useful for parents. The National Institute for Mental Health's Web site offers a comprehensive overview of depression, current therapeutic approaches, research, and suggestions on books to read.

162 **being unburdened by any tendency to overestimate themselves or their ability to control events:** Lauren B. Alloy and Lyn Y. Abramson, "The Judgment of Contingency in

Depressed and Nondepressed Students: Sadder but Wiser?"
Journal of Experimental Psychology 108 (1979): 441–85.

Chapter Seven: Coping

171 **Emily is a Presbyterian minister:** Emily is not her real
name; because her depressive episode began soon after she took
over the pastorate at her church, and she was worried that her
congregation would link her depression to her job, she asked
not to be identified in detail. The names of her children have
also been changed. Our interview took place on June 28, 2004.

175 **identifying a particular skill . . . and using it well is a
source of deep emotional satisfaction:** Ed Diener and Martin E. P. Seligman, "Very Happy People," *Psychological Science*
13 (2002): 81–84, cited in Martin Seligman, *Authentic Happiness* (Free Press, 2002), 42–43.

178 **"We're supposed to be superwomen—we handle everything with the greatest of ease":** Several women who corre-
sponded with me during the research for this book and who
identified themselves as African American (we did not ask for
racial identifiers in the survey) spoke of the special problems
they felt black mothers faced in admitting they suffer from depression. "Growing up in the African American community, seeing a therapist was taboo ('Oh, yeah, she a maniac'), not to
mention taking medication ('Oh, yeah, she really lost all her
scruples now')," wrote Sheila Whitesmith, a thirty-four-year-old
mother in Suisun City, California, who became suicidally depressed after the birth of her son, now seven.

185 **Richard Wagner, Salvador Dalí, or Gioacchino Rossini?:**
What's Opera, Doc?, Dough for the Do-Do, and *Rabbit of Seville,*
respectively.

188 **pretty evenly divided on this issue:** One survey question
asked, "Do you try to conceal your depression from your hus-

band/partner?" Out of 236 responses, 117 (49.6 percent) said, "No, not at all." Fourteen answered, "Yes, all the time," while the rest said that they sometimes tried to hide their feelings. "I ask my husband to help me. I tell him that I am not having a good day and I need his help," one mother wrote—and, as mentioned in chapter four, about 43 percent of the women said their husbands helped out "a lot"; the rest reported varying degrees of household help, sometimes unwilling.

188 **Spouses of depressed people are themselves at a higher risk of depression:** Nili R. Benazon and James Coyne, "Living with a Depressed Spouse," *Journal of Family Psychology* 14:1 (March 2000): 71–79.

190 **marijuana "exacerbated my depressed feelings":** Though there has been some research reporting that marijuana is helpful in the treatment of anxiety in HIV-positive patients, and inconclusive research on the relationship between mood disorders and marijuana use in adolescents, little or no research has been done on the use of marijuana among depressed adults.

196 **treatment for mental illness is far more expensive than treatment for other ailments:** The vast majority of insurance plans do not offer the same kind of coverage for psychiatric care as they do for non–mental health care, despite the passage by Congress of the Mental Health Parity Act of 1996. That law focused on benefits for "catastrophic" mental health crises and left it possible for insurers to offer parity on paper that does not translate to actual parity. Insurers may, for instance, advertise an across-the-board 90 percent reimbursement fee for all medical services, but that's 90 percent of a "usual and prevailing rate," a figure set by each insurer based on a variety of factors, none of which are known to the consumer. The "usual and prevailing rate" insurance companies set for mental health providers usually represents a much smaller percentage of the doctor's actual fee than is the case for other kinds of medical providers. Often, insurance companies re-

quire people with psychiatric issues to first visit their company's Employee Assistance Program. Not only are these programs staffed by people unequipped to deal with serious psychiatric issues, but the workplace stigma attached to being referred to an EAP effectively keeps many people from seeking psychiatric care.

201 **and some have really worked:** The term *natural* should not be confused with *safer*. Many so-called natural remedies are simply drugs that happen to occur in plant form, and they can interact with prescription drugs just like other prescription drugs do. Saint-John's-wort, for example, can enhance the level of some prescription antidepressants to the equivalent of unsafe dosage levels; DHEA (dehydroepiandrosterone) supplements are not recommended for anyone suffering from a hormone-sensitive cancer such as breast cancer. You should never take a "natural" supplement without telling your doctor.

201 **Nobody can quite explain how acupuncture works in terms known to Western medicine:** The National Institutes of Health Consensus Development Conference Statement on this subject, dated November 1997, reads in part: "Acupuncture as a therapeutic intervention is widely practiced in the United States. While there have been many studies of its potential usefulness, many of these studies provide equivocal results because of design, sample size, and other factors. The issue is further complicated by inherent difficulties in the use of appropriate controls, such as placebos and sham acupuncture groups. However, promising results have emerged, for example, showing efficacy of acupuncture in adult postoperative and chemotherapy nausea and vomiting and in postoperative dental pain. There are other situations such as addiction, stroke rehabilitation, headache, menstrual cramps, tennis elbow, fibromyalgia, myofascial pain, osteoarthritis, lower back pain, carpal tunnel syndrome, and asthma, in which acupuncture may be useful as an adjunct treatment or an acceptable alternative or be included in a comprehensive management program.

Further research is likely to uncover additional areas where acupuncture interventions will be useful."

Chapter Eight: How the Struggle with Depression Can Make You a Better Mother

210 **"Rather, [Lincoln's greatness] must be accounted an outgrowth of the same system that produced the suffering":** Joshua Wolf Shenk, *Lincoln's Melancholy: How Depression Challenged a President and Fueled His Greatness* (Houghton Mifflin, 2005).

212 **Nearly half of the mothers I talked to or heard from:** In response to the question, "In what ways, if any, has your depression helped you to be a better mother?" a total of 94 out of 231 respondents said it was by fostering their own sense of empathy. Another 22 said that the greatest benefit was that it helped them be more open in their communication. Thirty-one mothers said they saw no benefit at all from having suffered depression, and two others said they didn't see any connection, good or bad.

212 **and to seek help when they needed it:** In response to the same question as above, twenty-five mothers said that the biggest benefit in having been through the experience of depression was that it taught their children it was all right to ask for help; another seventeen said that their experience had educated them on the subject of depression and they had used that knowledge to help their children, or that they planned to; nine mothers said that it had broken a "code of silence" in their families about mental health topics.

216 **Denise Reid is forty-eight:** Interview, June 2, 2004.

Index

Index